W9-AFZ-627

AGAINST
BIOLOGICAL DETERMINISM

Also by the Dialectics of Biology Group:

Towards a Liberatory Biology

The conference on which the chapters in this book are based was held in Bressanone, Italy, on 26-30 March 1980.
It was organized by:
Martin Barker, Allan Muir and Steven Rose (in England);
Lauro Galzigna and Giacomo Gava (in Italy).
The conference papers were principally edited by:
Martin Barker, Lynda Birke, Allan Muir and Steven Rose with editorial advice and comment from Mae-Wan Ho and Hilary Rose.

AGAINST BIOLOGICAL DETERMINISM

The Dialectics of Biology Group

(General Editor, Steven Rose)

ALLISON & BUSBY
LONDON: NEW YORK

First published 1982 by
Allison and Busby Limited
6a Noel Street, London W1V 3RB, England
and distributed in the USA by
Schocken Books Inc
200 Madison Avenue, New York, N.Y. 10016

Copyright © 1982 by the Dialectics of Biology Group
All rights reserved

British Library Cataloguing in Publication Data
Against biological determinism.
1. Biology – Philosophy
1. Rose, Steven
574'.01   QH331

ISBN 0-85031-423-2
ISBN 0-85031-424-0 Pbk

Photoset by
E. B. Photosetting Ltd
Woodend Avenue, Speke, Liverpool.
Printed in Great Britain by
The Camelot Press Ltd, Southampton.

CONTENTS

Introduction

The Dialectics of Biology project

Some two years ago, one of us circulated to a few friends a project for a conference on "the dialectics of biology and society in the production of mind". The project was in the form of a manifesto, and began as follows:

"A strange fate has overcome traditional Western philosophy of mind. Over the past decades, its dominant paradigm has become the 'central-state materialism' of the identity theory, a hard-nosed reductionism whereby all mind events are seen as reducible to nothing but brain events to be interpreted by the triumphant progress of neurophysiology. The spirit which moved the brain and prevented it, in the old Cartesian system, from being mere mechanism, has all but vanished among philosophers. The historical reasons why the dominant ideology of bourgeois thought should have so inverted itself in recent years belong in the domain of the sociology of knowledge, and need study in their own right.

"What can be pointed to however, as a surely non-coincidental development, has been a similar hard-nosed biological reductionism which now plays a much wider ideological role. This mode of thought, somewhat grandiosely described as sociobiology, seeks and claims to find biological – even genetic or evolutionary – explanations for the widest range of social phenomena: wars, class structures, political and liberationist movements, criminality, intelligence . . . The phenomenal agenda of sociobiology has been widely proclaimed by its protagonists and as strongly refuted by many anthropologists, psychologists and biologists.

"Within the field of biology itself acceptance or rejection of reductionism has tended to separate along disciplinary lines. Molecular biologists in particular have been enthusiastic reductionists. Evolutionary, population and developmental biologists have not been swept along so far by this tide.

"What of neurobiologists and psychologists? Again there are indications of disciplinary divisions which may be of interest to the sociology of knowledge. Biochemists and pharmacologists tend towards reductionism; neurophysiologists and psychologists (except perhaps for black-box reductionists such as the Skinnerians) have been more circumspect.

"Reductionism as a research programme is fundamental to scientific method. Where it fails is in its attempt to function as a total explanation of the physical, biological and social universe, when it becomes an ideology, an epistemology that claims ontological status, either vulgar (as Sahlins describes it) or scientistic.

"Scientifically, therefore, reductionism leads only to paradox or impasse (as in today's theoretical physics). Yet ranged against reductionism appear to be mainly the forces of a reactionary and incoherent idealism, sometimes scientized into 'general systems theory' but predominantly searching for pre (or para-) scientific transcendence.

"Partly, this failure is indicative of the parochialism of a Western philosophical tradition which excludes from discussion any mention of that form of materialism which is non-reductionist, i.e. dialectical materialism; and tends to discount neuropsychology and neurophysiology conducted within that tradition. The political and ideological reasons which lie behind this exclusion are clear, and their consequences are profound. For such is the hegemonic power of the reductionist tradition in Western science (and philosophy) that even the new generation of critics of Western science's epistemology find it difficult to recover, or rediscover, the power of dialectical method. This is not to argue that the debased 'diamat' tradition of the Soviet Union is the way forward, or even that dialectics circumvents all theoretical problems; far from it. We do not solve anything by merely invoking the magic wand of dialectics or the concept of a hierarchical level. At best, we provide ourselves with a method of arguing and a conceptual framework within which – with the help of which, better – we can construct our arguments and order our observations. But despite the existence of – or our belief in the existence of – such a method, despite our need to assimilate the lessons of this dialectical tradition to our own critique, both of reductionism and of 'beyond reductionism' in the Koestlerian sense, we must recognize, I believe, that we must begin almost from the beginning in our attempts to understand what, for philosophy, neurobiology and perhaps sociology is the central question, the dialectics of biology and society in the production of mind.

"All of us know what we are trying to transcend. Several of us, who have discussed these ideas in twos and threes, partially and too briefly, may at best have a glimmering of what we are trying to create. Hence this proposal."

The Bressanone Conference

The response to this "manifesto" was the establishment of a small organizing group, including Allan Muir, Martin Barker (and until forced by pressure of work to withdraw, Tim Shallice) together with Steven Rose in England, and a generous offer from Lauro Galzigna to host the conference in a college of the University of Padua at Bressanone, from 26-30 March 1980, with the imposing but rather obscure title "The dialectics of biology and society in the production of mind".

We were anxious to keep the discussion viable by aiming for a relatively small gathering of forty to fifty people. This precluded mass advertisement of the event, so interest spread mainly through interpersonal contacts, resulting in an eventual attendance of fifty people. There were both strengths and weaknesses in this approach. On the positive side it helped to avoid the alienation resulting from a large, impersonal gathering since most participants were already a part of a friendship or professional network. More negatively, the international representation clustered around just a few countries; the strongest presence being from England, where the conference structure was largely planned, and from Italy, where the conference took place, with other participants from the USA, Australia, France and Belgium. We were thus in no way fully international. There was a conspicuous lack of attendance from "actually existing socialist countries" despite our invitations; indeed, all of us present were trained in the Western academic tradition and working in the institutions of advanced capitalist societies. Nor – a point that is returned to later – were we able to overcome the problems of male dominance in our organizing.

The disciplines – in so far as these are formal designations – included philosophy, sociology, mathematics, linguistics, psychoanalysis, psychology, ethology, pharmacology, neurobiology, developmental and evolutionary biology. The business of making it possible for people from these disparate backgrounds to communicate was achieved on the practical base of the warmth of the hospitality offered by the college at Bressanone and the organizational work of Lauro Galzigna and Giacomo Gava, together with the very serious intent of participants to talk and to listen through the disciplinary barriers.

The format of the meeting was to be as important as the content. To avoid long, formal presentations, people had been asked to circulate discussion papers in advance, and some twenty-five papers were received, either as background or introductory themes for discussion.

The papers in the two volumes *Against Biological Determinism* and *Towards a Liberatory Biology* were either submitted for discussion at the conference and revised in the light of discussion, or, in other cases, were written up after verbal presentation. The objective was to produce not merely the author's pre-conference thinking but to encourage an active response to the proceedings (although maybe this discouraged some possible participants by intimidating them!). The demands of preparing volumes of limited size while reflecting as many as possible of the multifarious papers and positions has made it necessary to shorten ruthlessly many of the papers – often much to the editorial group's regret. It is a reflection of the gentle and co-operative character of the meeting that no one protested at such treatment.

Although everyone at the conference agreed in their dissatisfaction with the prevailing neo-Darwinist paradigm and its concomitant reductionist ideology, there was no such ready consensus as to the principal cause for concern. As will perhaps be evident to the reader of these volumes, there were differing evaluations of what had been achieved within the paradigm, and further, the merits and deficiencies of reductionism as an essential strategy of science, and even the definition of the term, were at issue. The meeting was planned around six half-day topics:

1. Know your enemy – reductionism as meta-theory.
2. Systems, machines and dialectics.
3. Evolution, organism and environment.
4. Neurobiological explanations and human action.
5. The material basis of consciousness.
6. Where does our behaviour come from?

These sessions were anything but watertight, discussions from earlier sessions tending to spill over into later ones. Most particularly, the philosophical concerns of the first half-day – the significance of reductionism and the forms of its dialectical alternatives – were recurrent themes. Again, the reader will see from the present volumes the difficulties which authors felt in confining themselves within single-discipline questions – and we have reorganized the papers in a way which departs from the original sequence at the meeting, but now seems more appropriate.

Was the conference, then, a success? In that manifold issues and problems were raised and discussed in a way which broke discipline boundaries, yes; in that in general the feel of the meeting was warm and good, yes; in that coherent solutions were agreed, no. It *was* agreed, however, that the event was sufficiently worthwhile to merit a follow-up within a couple of years.

Any such enterprise will need to grapple with two major negative features of the Bressanone conference. Firstly, despite our efforts to avoid it happening, the conference participants fell into two groups, "insiders" who debated vigorously, and "outsiders" who were excluded from active participation by the meeting's processes – not least language. This exclusion was not merely linguistic, it fell, too, across sex lines, and was rapidly challenged by the women present, who organized themselves for part of the time into an autonomous discussion group. They pointed out that the organizing committee had consisted only of men, and that the style of the proceedings – certainly in the initial phase– made the participation of women difficult. This view and the response of the organizers are set out in the final section of *Towards a Liberatory Biology*.

Against Biological Determinism

The subsequent editing of the papers for the conference was entrusted to a small group of those who had been actively involved in the organizing of the meeting or of the conference itself. We wanted to produce material which conveyed the spirit of the conference while being accessible to the widest possible audience, and we have hence extensively edited and rearranged the material. In consultation with our publishers, Allison and Busby, we have separated the conference material into two volumes, each of some ten chapters, entitled *Against Biological Determinism* and *Towards a Liberatory Biology*. Each volume is "complete" but each complements the other. The first is primarily concerned with the philosophical, political and ideological challenge to biological reductionism, and its transcendence by way of systems or dialectically based theories; the second with exploring the building of a new biology – evolutionary, developmental and neurobiological – based on non-reductionist premises.

Against Biological Determinism begins with the theme "know your enemy". It asks: what do we mean by reductionism? The papers are not necessarily united in being "against" reductionism as empirical method, but are concerned to criticize both the current dominance of reductionist explanations within capitalist science, and specific instances of deterministic arguments within biology, medicine, and sociology. It is a belief common to us that there has been a recent intensification of deterministic explanations of human behaviour. These arguments provide support for views of the immutability and naturalness of behaviour, paying no attention to the form of the society of which the individual is part. More specifically, they lend support to the view that differences, whether between classes, races or sexes, are inevitable.

Several chapters discuss the historical emergence of reductionism as the dominant ideology of Western science, and in particular emphasize that reductionism, while initially liberatory, has now become oppressive. The first chapter, "Biology and Ideology: The Uses of Reductionism" by **Martin Barker**, sets out the major themes for discussion; the nature and historical development of reductionism, and its significance for our understanding of the relationship between science and ideology.

Janna Thompson, in "Human Nature and Social Explanation", points to the similar problem of sociological reductionism which, she argues, is just as much to be avoided as biological determinism. She refers to the need, in her view, to develop a theory of human nature, and suggests that the development of a truly dialectical understanding of the relations between the biological and the social demands that we take a general view of the relationship between human social needs and the natural world. **Hilary Rose** and **Steven Rose** ("On Oppositions to Reductionism") consider the recent history of reductionism and of radical critiques of it. They point to a tendency towards sociological relativism within many of these critiques, which they reject, and outline what they consider the tasks of a liberatory biology to be.

In the fourth chapter, "Cleaving the Mind: Speculations on Conceptual Dichotomies", **Lynda Birke** argues that reductionism is part of a tendency to categorize the world in particular ways, and uses as an example the tendency to root our conceptual models in rigid dichotomies. The dualities which pervade Western thought are often linked with rigid gender dichotomies, she suggests, which in turn determine what may be defined as "problems" within many areas of science and medicine. One kind of dichotomous classification is that between "normal" and "abnormal". **Lesley Rogers** in "The Ideology of Medicine" is concerned to understand the ideological role of Western medicine in creating and perpetuating this division. She also shows how reductionist accounts of the relationship between behaviour and biology can help to problematize the behaviour, thus rendering it a subject for medical focus. She analyses in particular the ways in which homosexual behaviour has become increasingly treated as a medical "problem", resulting in a variety of interventions.

In the last chapter of this group ("Disease Models and Reductionist Thinking in the Biomedical Sciences") **Giorgio Bignami** discusses further the ideological function of medicine, and dwells particularly upon post hoc explanations of disease states within the biomedical sciences. At its most extreme, this may take the form of diseases becoming defined simply in terms of the drugs which are known to relieve their symptoms; hyperactivity in children, for example, becoming "hypo-amphetaminosis".

The second group of papers of this volume begins the attempt at transcending reductionist thinking. **Giacomo Gava** ("Hierarchical Structures and Structural Descriptions") takes up the idea of different "levels" of language, emphasizing the hierarchical structure of such levels and the untranslatability between languages of scientific discourse, each of which operates within its own logic.

Chapters 8 and 9 are concerned with the distinction between matter and information. Each deals, in these terms, with relations between levels of description in complex systems, emphasizing the role of openness in the establishment of order. **Allan Muir** ("Holism and Reductionism are Compatible") argues that the matter/information distinction is the key to overcoming the familiar disjunction between reductionist and holistic accounts of phenomena. Following this, a discursive survey of the same problems in the theory of knowledge production is designed to show how the information concept can help to resolve other major dichotomies.

The paper by **Lauro Galzigna,** "Matter, Information and their Interaction in Memory Processes", draws upon the theory of co-operative models in the physical sciences. Statistical treatment of an ensemble of non-linearly interacting system components reveals how large-scale ordered behaviour can result from disordered phenomena at lower levels. Such models of self-organization in thermodynamically open systems can be interpreted as information registration, which, in turn, indicates how a theory of biological memory can be approached.

The final paper of this volume, "Reductionism Reassessed", by **Werner Callebaut,** is an attempt to point the way beyond reductionism through a detailed examination of the logical positivist programme for establishing the "unity of science". Connections between reductionism, the conceptual or perceptual simplification of complex systems and relations between theories are discussed. It is suggested that the overcoming of reductionism in theory may be connected with the re-shaping of science as a social institution into a non-hierarchical form.

1
Biology and Ideology: The Uses of Reductionism
Martin Barker

It is a matter of more than passing interest that reductionist biology has been such a source of support for reactionary political views. Why should this be so? We cannot be satisfied with an answer solely in terms of misuses of otherwise innocent biological investigations. What then is the basis for the ideological significance of reductionist biology?

I pose the question in this way because of something Steven Rose has said. In dismissing (surely rightly) the "science as social relations" school, he commented:

> To understand the significance of the current revival of biological determinism it is necessary to examine both its social and political history and the truth content of its present claims. [1]

The trouble is that this, while true, leaves unclear what is to be the relation between investigating the ideological nature of these ideas, and testing their truth. Indeed I find it very significant that there is a long tradition of asserting that the two enterprises are quite separate. Where writers have wanted to use a concept of ideology, very often they have put it in terms of the "interests" or "values" of scientists interfering with the objectivity of their science. Or they have fallen into discussing the "social determinants" of ideas. These approaches have in common the assumption of a clear distinction between truth and ideology, and, which is particularly important, both assume that there are definite senses in which ideology is non-rational, determined by non-rational processes.

It seems to me that it is precisely this view that underlies the explanation most commonly offered for the ideological import of reductionist biology. Erich Fromm expressed it perfectly when he wrote:

> Perhaps Lorenz's neo-instinctivism was so successful not because his arguments are strong, but because people are so susceptible to them. What could be more welcome to people who are frightened and feel

impotent to change the course leading to destruction, than a theory that assures us that violence stems from our animal nature, from an ungovernable drive for aggression, and that the best we can do, as Lorenz asserts, is to understand the law of evolution that accounts for the power of the drive?[2]

Notice the wording of Fromm's case (repeated, incidentally, almost word for word in Leakey and Lewin). The arguments are weak, but people are susceptible. In other words, Lorenz has appealed to something other than reason.

This puts Fromm, and all who agree with him, firmly within a tradition. It is a tradition that spans back to Francis Bacon and John Locke; informs David Hume; and is to be found to this day in the ideas of Popper, Feuer, Naess, and a host of others. This tradition seeks separate determinants for truth and (ideological) falsehood. "Truth" is got by using correct procedures, proper scientific method; ideological falsehood is got by letting something else interfere. David Bloor characterizes this dualism very aptly:

> When men behave rationally or logically it is tempting to say that their actions are governed by the requirements of reasonableness or logic. The explanation of why a man draws the conclusion he does from a set of premises may appear to reside in the principles of logical inference themselves. Logic, it may seem, constitutes a set of connections between premises and conclusions and men's minds can trace out these connections. As long as they are being reasonable, then the connections themselves would seem to provide the best explanation for the beliefs of the reasoner. Like an engine on the rails, the rails themselves dictate where it will go . . . Of course, where men make mistakes in their reasoning or logic, then logic itself is no explanation. A lapse or deviation may be due to the interference of a whole variety of factors . . . As when a train goes off the rails, a cause for the accident can surely be found. But we neither have, nor need, commissions of enquiry into why accidents do not happen.[3]

It is this (empiricist) tradition that warrants all the talk of the "values" or "interests" of scientists interfering with their work, or of the views of a scientist being determined by political, social or historical factors.

I have a host of objections to this account, for most of which I lack space.

Fromm's statement, as a particular expression of this general approach, encourages lazy investigation. For it is not only Lorenzian ethology that has theorised "aggression" in terms of evolutionary functions. Sociobiology has, too. But there are big differences between the particular answers they come up with; Fromm's account does not allow us to differentiate them. There is a great deal of evidence that views such as Lorenz's, sharing all the characteristics of suggesting inevitability of biological responses, have been and are being used to *change* social and political relations (or what people often dangerously call the "status quo"). Allan Chase gives many horrific descriptions of the actions of the eugenicists in the United States who pressed, successfully and self-consciously, for the application of social Darwinist ideas. Coming up to date, we have already seen the beginnings of sociobiology being put to use in support of the Thatcherite backlash in Britain, and of the anti-feminist backlash in America. And I have deliberately avoided mention of Nazism and associated fascisms. *These ideas are weapons in the hands of activists,* not, as Fromm suggests, excuses in the hands of cowards.

Lastly, and perhaps to me most importantly, I want to suggest that this empiricist opposition between truth and ideology is actually part of the very syndrome of ideas that has given credence to biological determinism. It is not possible to do much more than illustrate this point here. I have developed it at length as a case-study of David Hume's racism, in my forthcoming book on racism. And it is surely important in itself that the same tradition that provided the major arguments in favour of reductionism, via the empiricist atomization of experience, is also the one that built so much on the division between reason and the passions. But to illustrate where this has led, let me quote an early twentieth-century social Darwinist, William Trotter. Trotter and his contemporaries were puzzled by something that our recent reductionists have neatly forgotten: what does it feel like to have an instinct? This was something of real concern to Trotter, MacDougall, Rivers and many others at this period. Trotter's answer was that, because instincts were so ineluctable, they must present themselves to the mind as "a priori syntheses of the most perfect kind"; that is, as beliefs which were unquestionable, ideas which it would be absurd to doubt. Here he presents an opposition that is easily recognizable:

> Non-rational judgements, being the product of suggestion, will have the quality of instinctive opinion, or, as we may call it, of belief in the strict sense . . . When therefore we find ourselves entertaining an opinion about the basis of which there is a quality of feeling which tells us that to inquire into it would be absurd, obviously unnecessary, unprofitable, undesirable, bad form, or wicked, we may know that the opinion is a non-rational one, and probably, therefore, founded on inadequate evidence.

> Opinions, on the other hand, which are acquired as the result of experience alone do not possess this quality of primary certitude. They are true in the sense of being verifiable, but they are unaccompanied by that profound feeling of truth which belief possesses, and therefore we have no sense of reluctance in admitting inquiry into them.[4]

Those two opposed sources of beliefs again– only this time beliefs that are weak on argument but strong on commitment are rooted in our biological make-up. And that concept of "suggestion" is found all the way back to Hume, amounting to a sort of emotional contagion rooted in our "original nature".

A vast quantity of discussion of ideology is tainted with this distinction. It is, indeed, the source of the all-too-common and often-too-correct critique of marxists that they claim everybody else's beliefs are socially determined, but appear to exempt their own. And here "social determination" has the force of countermanding claims to rationality and truth, of indicating origins in "interests", in the "reflection of material relations in ideas", and so on.

In order to begin the question of ideology in a way that hopefully avoids the traps I have suggested, I want to focus on one point in sociobiological thinking. This is its dependence on the idea of genetic units. Perhaps the most "charming" example of this is in Robert Trivers's classic article on reciprocal altruism.[5] He takes this to cover equally the symbiotic relation of cleaner fish and their hosts, alarm calls among flocking birds, and humans being nice to each other. Properly to sense its form his account should be read in full. Note here only the central use of the gene-unit idea:

> Assume that the altruistic behaviour of an altruist is controlled by an allele (dominant or recessive) a^2, at a given locus and that (for simplicity) there is only one alternative allele, a^1, at that locus and that it does not lead to altruistic behaviour.[6]

Genetically, there are numerous reasons to regard this as absurd. But just why are the sociobiologists tied to the idea? I want to explore this through an extended idea.

Richard Dawkins's book *The Selfish Gene* has the advantage of being honest enough to display its errors. Dawkins begins with an attempt to explain what he means by genetic selfishness:

> An entity, such as a baboon, is said to be altruistic if it behaves in such a way as to increase another such entity's welfare at the expense of its own. Selfish behaviour has exactly the opposite effect. "Welfare" is defined as "chances of survival", even if the effect on actual life and death prospects

is so small as to seem negligible . . . It is important to realize that the above definitions of altruism and selfishness are *behavioural*, not subjective.[7]

For all its oddity, this is at least clear. But it is disturbing how quickly this apparent clarity vanishes. By page 10 Dawkins is talking of trade unionism as an example of altruism within a group, but selfishness between groups. Presumably he isn't seriously proposing that trade unionism is to be understood in terms of enhancing survival and reproductive chances. And then in chapter 3 we find that a really crucial transition in meanings has taken place – to talking of genes as the units of selfishness:

> We have now arrived back at the point we left at the end of chapter 1. There we saw that selfishness is to be expected in any entity which deserves the title of a basic unit of natural selection. We saw that some people regard the species as the unit of material selection, others the population or group within the species, and yet others the individual. I said that I preferred to think of the gene as the fundamental unit of self-interest. What I have now done is to *define* the gene in such a way that I cannot really help being right![8]

How has he managed this astonishing feat? It is instructive to retrace his argument. After discussing the nature and transmission of genetic information, he discusses the formation of genetic units. Consider mimicry, he invites us, whereby the hoverfly for example gains protection for itself by copying the morphology and behaviour of wasps. In what does mimicry consist? It comprises, at least, striped coloration, pointed tail (indicating potential sting), and pattern of flight. For mimicry to work, it is essential that the required characteristics are all present. But that implies, says Dawkins, that they have been "grouped" genetically: a process of evolutionary editing must have gone on such that these characteristics now form a genetic unit, for all practical purposes. It is the unit of mimicry.

This is a very revealing argument. For in the very logic of his account, evolutionary strategies have to be located first; only then can we infer a unitary genetic cause. But that assumes that all behaviour always comes thus grouped into units; and we can discover the nature of the unit provided we are prepared to go on looking long enough for the underlying unity and commonality. I am putting it thus because I want it to be seen that the form that sociobiology takes is of a *project of innateness*. It is not so much a theory as a proposal or invitation. For it merely says that there must always be such units; it doesn't tell us how to find them.

The uses of this can be seen through an example. David Barash considers the significance of the evidence for inheritance of intelligence; he describes the Tolman and Tryon experiments on the selective breeding of maze-bright and maze-dull rats:

> These two strains were maintained separately and their descendants were tested again in the original maze, many generations later. They still retained their bright and dull performances, respectively. But when tested in a maze that was slightly different from Tryon's original one, there was no significant difference between the two populations. The implications of this finding are obscure, but it suggests the caution that selection for behaviour may actually involve finer and more precise distinctions than we, as selectors, are aware. [9]

Given difficulties with the evidence, we must go for finer distinctions, smaller units than before. The project of innateness will not allow us to go in the opposite direction, towards denying that intelligence or any subcomponent is the sort of thing for which it makes sense to seek direct, quantifiable genetic transmission.

Seeing sociobiology as a project of innateness has several major advantages that I want to take in turn. First, it reveals the mechanism of penetration of ideology into science. Second, it tells much about the nature and motives of the sort of reductionism with which we are concerned. Finally, therefore, it hints at what we need as a replacement account.

Sociobiology requires us to break behaviour into units, even if it resists, for purposes of explanation. But it cannot provide us with methods for determining the content of these units, other than by hypothesizing that some bit of behaviour constitutes an evolutionary strategy. It is just here that we can locate the ideological mechanism. Not having an independent method of determining the nature of the units, sociobiology is vulnerable to the forms of "common sense" which tend to identify unitary proximate causes within such phrases as "It's only natural to . . ."

I believe this enables us to grasp much about the particular accent of sociobiology. Consider its main claims to fame: its account of kin-selection; its account of male and female strategies; its explanation of racial prejudice in terms of the boundaries of genetic relatedness; its picture of aggression. Each of these maps with considerable accuracy on to aspects of contemporary reactionary ideology – the ideology of the family; the reaction against the women's movement; the "new racism" that asserts the naturalness of xenophobia while explicitly disavowing assertions of superiority; and the current concern about "violence", conceived as aggressiveness that has lost the balances provided by adequate social controls.

This last can provide a useful test-case of the approach I am taking. Earlier one of my objections to Fromm's accounting with instinctivism was that there would be a tendency to gloss the differences among the reductionists. I think that the approach I am taking can cope both with the commonalities and the differences between sociobiology, human ethology, MacDougall's instinctivist psychology, etc. Now the ethological picture of aggression, of its ritualization and redirection, does not match the sociobiologist's. In a way, the ethological approach is suggesting that certain forms of violence may in fact be socially useful; for they fulfil evolutionary functions of territory-maintenance, of making social hierarchies and so on. And far from inviting social controls on violence, Lorenzian analysis requires that we find increased ritual outlets for aggression.

We can understand this view of aggression if we look at how the ethologists conceptualized what we may call their "founding problem". Tinbergen gave particularly apt expression to it:

> Man is the only species that is a mass murderer, the only misfit in his own society. Why should this be so?[10]

The context in which human ethology flowered was after the Second World War, and the key outcome of Lorenz's and others' account of ritualization of aggression was an explanation of nationalism. Commentators have rightly drawn attention to the importance in Lorenz's theory of the notion of "militant enthusiasm". This was the core of his account of the instinctive expression of aggression as nationalism; it was genetically driven, socially guided aggression. It is not hard from all this to understand the ethological obsession with aggression, and the form of their explanations of it, as a partial guilt-response in the aftermath of Nazism. Their attempt at a Darwinian explanation of aggression, of course, assumed a historical picking out of aggression, mass murder, and militant enthusiasm as "unitary" problems needing an answer.

The contrast with sociobiology is very striking. Indeed, many of the sociobiologists have made their different view of aggression one of the main points of separation. For them, aggression is a natural relation even between members of the same species. Each organism calculates its own advantage (blindly, of course, but that never seems to make any difference). That does not mean that all organisms must be aggressive to one another. There is a balancing process inscribed in the evolutionarily stable strategy (or ESS) that, *other things being equal*, there should be no excess, i.e. damaging, aggression within a species. But what is society but an ESS in action? The picture of aggression that results from this fits nicely with political discussions on the need, nay the right, of people to be restrained within a social collective.[11] And violence when it occurs will be seen as the result of other things becoming unequal.

This only covers part of the problem of ideology presented by reductionist biology. It shows how commonsense political ideas get appropriated by science, and therefore are returned to the world with a spurious objectivity. But into what are such ideas appropriated? For there are very important differences in the ways in which evolutionary mechanisms have been pictured. Jill Quadagno has shown the extent to which sociobiology has grown on an assumption that all organisms display *maximum adaptiveness*.[12] She points out that, by its nature, this idea is in principle untestable because, except under extraordinary circumstances, it contains the necessary assumption that all organisms have been selected that will maximize their fitness. Indeed without that assumption the ESS concept would be empty.

It is interesting to speculate about the political parallels of this view. For it neatly corresponds with a view of capitalism as natural and endemic in the nature of all living things, and of the market (as the pool of competing strategies) as the selector of fitness. But for all the suggestiveness of this, it is as well to consider how the sociobiologists have themselves justified the idea of fitness-maximization.

This is in fact not an issue on which, so far as I know, they have been particularly forthcoming. But one can extract from their general arguments what their answer would be. There seem to be two main components: one is an argument from continuity; the other is an argument from "genetic cheating". Both take us into the heartland of reductionism; both subscribe to the unitization of behaviour.

The continuity argument, put baldly, is not a bad one. It demands that we see humans as evolved organisms. But of course it only has substantive content as a demand if we can say interesting things about the other species with which we are to be judged continuous. Here is where the reductionism shows. For the classic argument has been that we know that non-human species' behaviour has genetic bases; therefore human behaviour must have. Again, one can sense the ambiguities underlying the phrase "genetic determinism"; for in one good sense it must be true. Given that the human organism is a structure of proteins and other biochemical combinations built by genes, anything we do is genetically determined. The interesting, but also false, part of their claim is that each mode of behaviour has a *separate, and therefore* potentially detachable, genetic determinant.

Dawkins provides us with a good example of this claim about unitary determination. He makes much play of the experiments on hybridization of normal and "hygienic" bees. If valid, the experiment would seem to show that hygienic behaviour among honeybees is governed by two separate genetic components which are theoretically, and under experimental conditions practically, separable.[13]

The sociobiologists have argued strongly (and rightly) that genetic mutation effectively has to take place one gene at a time. The mathematical chances of two simultaneous mutations occurring, and proving non-deleterious to the organism, are infinitesimally small. Doesn't this prove, say the sociobiologists, that individual behaviours have individual determinants? For each difference in behaviour-potential, there is a specific gene-difference. What is wrong with this as an argument? Add in the genetic cheat argument. Simply, the sociobiologists argue that if we suppose a population living below its maximum use of its environment for reproductive purposes, the chances are that a deviant individual will occur which will take advantage of the "slack" left. On average, such an individual will tend to leave more offspring. Therefore the genes for more complete use of an environment will tend to spread and to dominate the gene-pool. How should we rate this as an argument? Taken in conjunction with previous arguments doesn't it tend to show that all behaviour is determined in units, and that the units present in species actually existent must approximate to maximum (selfish) fitness?

Things like this can surely happen perfectly well. There will be cases where a difference in behaviour and/or morphology can be traced to a unitary difference in the gene-base. Eye-colour is such an example, as I understand it. At the moment I see no reason to suppose that the bee experiments have not revealed another such example. But there is a twofold error in generalizing from this to a universal principle: first, by no means all gene variations make no difference to other genes around them. Secondly, it invites us to ignore the specific character, or content, of what is thus generated by the genes.

It has always struck me as ironic that both sides in the argument about reductionism regularly resort to the same paradigm example: phenylketonuria (PKU). This has been used by reductionists as proof that human behaviour has genetic preconditions (and therefore is genetically determined, they would say): it has been used by their opponents who have pointed out that the gene-difference loses its effect (chronic brain-damage) if the diet is varied or if injections of the missing enzyme are given (environmental alteration). But surely the interesting facts about PKU are that we have the ability to investigate, understand, diagnose, and intervene in the situation (something no other species has yet managed). Furthermore, we intervene with the intention of enabling children thus afflicted to grow up displaying what we regard as specifically human attributes, which would be absent if brain damage was allowed to occur (ability to learn, develop, play, act, enjoy, etc.). But here I want to use the example for another purpose. The important thing about PKU is not that it results from a particular gene-difference, but that this gene-difference, and the enzyme lack that results from it, *interferes with the action of so many other genes*. PKU is a bloody nuisance to the whole gene-structure, especially to those parts that are at work on the brain. True, the heart is not particularly bothered by PKU, nor are other parts of the auton-omic nervous system. But here is a case of a gene-difference which cannot be held separate; it interferes pretty well across the board.

It seems to me that this shows up something very wrong with the genetic cheat argument. For it assumes that the cheat – with its new bit of behaviour – can have the new bit without affecting the function-ing of all other bits in any significant way. Now, it may sometimes be true (and as I have said, perhaps the bee example is one such): but without exploring the concrete content of a case, you cannot say. *And you cannot possibly build an a priori theory of maximization on the premise of the possibility of cheating*.

Despite its illegality, this move is regularly made by the sociobiologists, and it is in this, surely, that the second aspect of their ideological import lies. Sociobiology's specific politics reside in its absorption and transformation of the cultural units which it requires to complete its account. But the particular mechanism into which they are taken, and which Darwinizes them, has its own general political character. In this, individuals are conceived as necessarily having a profit-and-loss relationship with the rest of the world. Individuals develop orientations very like computers, operating input and output accounts with the environment – and the environment includes all other members of the species. The only test of success is survival. Also, there is a postulated relationship between the whole individual organism and its part-characteristics. For each characteristic is accounted for in the whole strategy of the organism, to be deployed according to calculations of efficacy and advantage. Each behaviour, with its potential generated by a separate determinant, becomes an *investment* (literally the sociobiological term), a commodity with an exchangeable worth in the market called "environment".

I haven't the slightest doubt that the sociobiologists would cry "foul" at such an account, and would claim that this is just the sort of sneer that was thrown at Darwin, but which really applies only to Spencer. (See the opening pages of Barash[14] for a case of this.) But I suspect that Darwinism always *was* political in a definite sense, and I think that the role that the conceived mechanism has within this latest variant of the reductionist strand within Darwinism entitles us to see it as political. It is a *contentful* proposal to research, disguised as a *formal* account.

To see the force of this argument, it is necessary to turn to the second criticism I had of the generalization proposal. It is here that we reach the core of reductionism, and of the weakness of many responses to it. As I have already suggested, we cannot decide in advance of looking at a specific gene and its "effect", whether it will retain in life its separateness. To me, this implies that the question of reductionism is not just, nor even primarily, a methodological question.

What is reductionism? According to most accounts, it is the epistemological proposal and practice of seeking explanatory accounts of phenomena in terms of some fundamental principles, in which complex phenomena are understood in terms of basic parts out of which they are composed. I don't believe it. This is a *part*, but *not the crucial part*, of reductionism. Missing is an answer to the question: what happens to the complex phenomena with which we began, which had to be "reduced", when we reach the reductionist explanation? The answer surely is, that they are done away with. This can of course be restricted to a theoretical level. If we seek to understand children in a school shouting and moving about in terms of "hyperactivity" resulting from brain malfunctions, then the language of shouting and moving about is to be done away with. It is got rid of, in favour of the language of hyperactivity. But I believe that reductionism is only comprehensible ideologically if we see that we cannot maintain a theory/practice division here. The recommendation to get rid of the language that accepts the complexity of the phenomenon can also be a practical political recommendation to get rid of the phenomenon itself.

The most common form of opposition to reductionism is to assert that there are different levels of explanation and there is not a most basic one that has priority over the rest. Steven Rose expresses this view in his *The Conscious Brain*:

> There are many levels at which one can describe the behaviour of the brain. One can describe the quantum structure of the atoms, or the molecular properties of the chemicals which compose it; the electromicrographic appearance of the individual cells within it; the behaviour of its neurons as an interacting system; the evolutionary or developmental history of these neurons as a changing pattern in time; the behavioural response of the individual human whose brain is under discussion; the familial or social environment of that human, and so on. Each of these descriptions may be complete in its own terms, yet which one is relevant must depend on the circumstances.[15]

Rose quite rightly criticizes any move from correlation between levels of analysis, to causal influence. At a negative level, and for purposes of particular research, this is fine. But it raises a number of awkward questions.

It is not clear on this view whether we are ever entitled to *prefer* one level of interpretation to another. If there are merely translation-rules between levels of explanation, and the choice of level at which to operate is a question of "whether the purpose for which the description is intended is related to the writing of a novel, sociological analysis, chemical description or experimentation on cellular phenomena" (ibid), then we are never entitled to require that one level of analysis has preference and precedence over others. That is a problem, since it does seem to me that at opposite ends of the spectrum, PKU and "hyperactivity" do have preferred explanations. The problem in the quotation is that it does not distinguish sharply enough between description and explanation. Undoubtedly an account of many sociological problems could be sharpened by a description of their biological components. Obvious cases such as drug addiction spring to mind. But that is not the same as saying that the biochemical analysis explains phenomenal expressions "at other levels". The question of explanation is left untouched. But the tempting way out, to deny that science deals in explanation in such forms, exactly reproduces the position I noted above, that there is no way of choosing a preferred level. Such methodological liberalism, looking nice and undogmatic, in fact seems to provide respectability for modes of intervention (e.g. drugging of schoolchildren) on the grounds that, scientifically, nothing can be said against them. Even if it hasn't worked yet, that is merely an invitation to further experiments.

Secondly, an epistemological account of levels has no way of answering the question: how many levels in each situation? This question does not imply an intention to reify. As noted before, the particular break between molecular biology and biochemistry, for example, is partly a historically induced distinction. But surely there are more types of things to be said about (a) a living organism of a simple kind, compared with a non-living physico-chemical structure, (b) an animal with motile behaviour, compared with a plant whose reactions, while complex, approximate more to reflexes, and (c) a self-conscious species (us) compared with a species in which genetic determination is relatively unmediated. By what rules do we admit new "levels", and what weight can they be given in an analysis?

We need to look at reductionism differently. I think an argument can be made that Darwinism, since its foundation, has contained the potential for two incompatible, but nevertheless coincidental, ontologies: that is, general theories of the nature of living organisms. The first, reductionist, strand in the tradition has offered what amounts to a *mechanism in general of the non-rationality of behaviour.* Looking again at early social Darwinism, it is very striking how strong is a myth among its practitioners that Darwinism represented a sharp break. Prior to Darwin, there was a dominant world-view that while "brute beasts" were mechanical and blind, humans were rational. Comes CD, so the myth runs, and this was challenged. Listen to how clearly one propagandist presented this opposition:

> The instincts, as inheritances from the past, have been the subject of investigation for centuries, but until the last quarter of a century they have been overshadowed by the attempt to prove that different from the animals, man was controlled by reason and not by instincts. When it was declared that man had more instincts than the animals, a stir was noticed among some scholars; but a score of years in the present century has done more to show the dependence of mental states upon past racial experiences than all the years before.[16]

The rewriting of the past implicit in this is important. For it shows that the theory of evolution was taken by many of its early devotees as an account of the non-rational sources, *and therefore present nature,* of all behaviour. In fact, an enormous number of pre-Darwinian thinkers had subscribed to this view. In the English tradition, empiricism was long touched with this view; and in David Hume and Adam Smith, it was made into a coherent theory. In the French Enlightenment, the whole tradition of "man a machine" broke the non-continuity of human and animal. In Germany, the romantic tradition had no time for notions of human rationality. But the myth played a role in aiding the capture of Darwinism as a theory of a general mechanism of non-rationality.

The alternative account that seems to me implicit in Darwinism is also an ontological account. It amounts to an assertion that, having evolved by natural selection, by selective survival in the face of environmental pressures, the characteristics species display are relevant to the modes of environmental pressure exerted on them. I call this the beginning of an ontology because, unlike social Darwinisms that have focused on *processes* of selection, this approach starts from the concrete nature of the evolved responses. Spencer by contrast made play with the phrase "survival of the fittest". Its ideological nature lay not only in its congruence with certain needs of capitalism, but in its reduction of the specificity of human beings to a general evolutionary process and mechanism. McDougall and his co-workers developed a process-theory in which "higher" and "lower" became evolutionary categories.[17] The ideological significance of that lay not only in its racist implications, but in making "higher" and "lower" functions of external processes, according to which societies could be graded. The alternative is already written in as a possibility in the main concepts. For example, consider one central term of the Darwinian syllogism: adaptation. Behaviour is adaptive in so far as it has evolved by selective survival within a population, and the environmental constraints that selected it remain relatively constant.

The concept of adaptation in fact contains many unclarities and ambiguities. Most important for my purpose is that between being adaptive and being adaptable. What sort of a distinction is this? Species with hard-wired behaviour relevant to environmental constraints can be said to be adaptive. Being hard-wired, their behaviour is relatively constant. But it is viable in evolutionary terms. There are, however, considerable difficulties about giving even a description of adaptability that is adequate. And yet it is not at all difficult to deduce its possibility from the central propositions of evolutionary theory, about natural selection. For if a species survives because its behaviour has taken account of its environmental situation, then one that can learn from its environment clearly can develop ways of behaving favourable to its survival.

But to describe adaptability is not the same as describing any particular behaviour. The behaviour that occurs at any time is a sort of fixing of what is otherwise a potential. The overt behaviour can therefore appear rigid and hard-wired, whereas what I would call the evolutionary strategy of flexibility that prompts it has selected it from an indefinite range of possibilities. I put it thus because I think it shows that this whole strategy of survival, completely comprehensible in terms of the core concepts of evolutionary theory, necessarily avoids the eye of the theorist who begins by looking at evolution as a general *process*. For in such a perspective, what is examined is behaviour, to see in what ways it is or is not adaptive.[18]

To describe an organism as adaptable is to introduce a mediating factor between the organism as an example of genotype, and its behaviour. Accounting for the behaviour is not the same as characterizing the organism. Or as Marx put it:

> The animal is immediately one with its vital activity. It is not distinct from it. Man makes his vital activity into an object of his will and consciousness. He is not immediately identical with any one of his characterizations.[19]

Marx worked with far too hard an opposition between humans and animals. To a considerable extent, many other species have evolved strategies of adaptability. But the insight was significant nevertheless.

The mark of adaptability is the ability to take account of an indefinite number of features in one's environment. This obviously involves learning. An interesting point about learning is the different forms it takes. There is, first, the sort of learning that, for example, Lorenz has been keen to point up: learning as the method of completion of already established goals or processes. The wasp that learns the geography of its nest-area is performing a fixed activity, whose completion will not alter the nest-making activity. But in so far as a strategy of adaptability is conceivable, learning has to go much further. What if the wasp could decide to relocate its nest after considering the terrain? What if it could measure wind, rainfall, chances of erosion or flooding? What if it could develop new building materials? Or if it decided that it preferred a communal nest, if previously solitary; or vice-versa? Logically, a strategy of adaptability could include such things. And, of course, it is behaviours like this that humans display.

The logical outcome of this is that we cannot talk of genetic determinism here other than as a truism. First, learning can only be flexible if it is not already "set" on to objects of interest. To be fixed is to be adaptive. Second, within the range of environmental features which could be made relevant by such a strategy comes the organism itself. Its body, its powers, and its intentions themselves become variables within the organism's planning of its life. Of course there are an enormous number of constraints on the plans and purposes of an adaptable species, including genetic and other biological constraints; but the significant thing is that they are capable of investigation, adaptation and alteration. This is once again the significance of PKU, as I suggested earlier. To the extent that we conquer for a child its tendency to brain damage, we enable it to live by its potential for adaptability. PKU is a constraint to be overcome in the organism, so that the strategy of adaptability can be followed.

The culmination of such a strategy must be a species in which all aspects of the environment are open to investigation for their potential relevance to pursuing plans, and plan-formation. The latter must be included, for inscribed in such a strategy is that such organisms must plan their own lives. They must make their own natures.

These points begin, I hope, to make clear how Darwinism can have within it a strain towards an ontological account of living beings. *A living being is that which has the capacity to make an environment for itself, and to maintain it to some extent.*

We can see evolution as the making and testing of modes of control of environment; and the organism as an environment in itself, or better, a part of its own environment. In fact, the nature and degree of connection/separation between an organism and its environment is a major part of the ontological equation. I do not mean only in the ecologists' sense, that food, for example, must enter the organism from the environment, and waste products leave it to be assimilated elsewhere. I mean also the ways in which an organism might be open to, or insulated from – or open itself to, or insulate itself from – environmental pressures and processes.[20] This is a sort of ontology of interaction and change. It describes, I think, some of the fundamental characteristics of living structures, and the levels at which they may be understood and investigated. Philosophers might complain that it is not an ontology in the traditional sense, in not giving one or more most basic types of existents. But to accept that criticism would involve assuming that ontology is what is reached after we have completed the reductionist procedures. Commitment to an alternative seems essential as part of an adequate response to reductionism. I suggest that an ontological account built on the ideas outlined above could be of help. If it has any validity, it has the virtue of bypassing all talk of levels, since it is concerned with the specificity of a species's response to its environment(s).

Such an ontology is necessary because of the challenge from

process-ontologies, that is, general theoretical accounts in which the actual species which result from selection pressures are reduced to functions of those processes. The general adherence to such process-ontologies surely constitutes a third level of ideological significance, precisely in that they demand that the concrete nature of any evolutionary product be re-presented as a function of an abstract process. The match with Marx's presentation of the subsumption of use-values under the commodity-form is, to me, very striking. That is not a proof, but a proposal. It would be interesting to see a reworking of the history of social Darwinism which explored the three dimensions of ideology which I have suggested. First, the most general, the adherence to a process-ontology; the sources, forms and changes in this. Second, the specification of evolutionary mechanisms, and the central concepts within which evolutionary theory is organized. I have already suggested that there may be relatively distinct periods, which may well correspond with broader political and ideological movements. And finally, there are the particular units that get taken up and made into evolutionary functions, strategic behaviour units against which (in particular, human) behaviour is then measured.

Having journeyed thus, it is possible to go back to the question of ideology. If we look at the position which beliefs have in the account of humanity I have sketched, they are part of the learning and planning process. Without developing the case in full – I am not at all sure I am able to – what is significant about a pattern of belief such as reductionism in general, or sociobiology in particular, is that it is not so much a belief about the nature of things, as a proposal (a) how we should examine the world, and (b) a recommendation as to how we should treat it, including ourselves. Now what is important about the strategy of adaptability is that it uses investigations of the world, selecting and ordering features felt to be relevant, in order to form plans. But recall that there is a sense in which an adaptable organism is not identical with any of its particular modes of behaving. It makes itself like that by committing itself to that way of living. Or, more aphoristically, "humans make themselves". To make ourselves is to form plans, and in the living out of them to fashion or refashion our beliefs, our consciousness, our social relations and our motives. For most of our history, this is done blindly. There is, to borrow a phrase, a gap between theory and practice.

Earlier I argued that if reductionism is understood epistemologically, we are left not knowing what happens to the phenomena that are reduced. Sociobiology, as earlier versions of reductionist biology, knows what to do. It operates with a distinction between those "natural" processes which are suitable for biological explanation, and the "artificial" elements and conditions created by culture, civilization and technology.[21] It is in fact a recipe, not for understanding the world only (albeit in reduced ways), but for changing it. Thus could the "artificial" phenomena be got rid of. Isn't this the moral underlying the use by the fascist National Front in Britain of such ideas? For all their confusion between sociobiology, ethology, and earlier social Darwinism represented on its last legs by Darlington, about the practical consequences they are quite clear:

> It has been said by Professor Darlington that man's faculty of reasoning has not destroyed the instinctive basis of his behaviour, but it has masked or distorted it. The great question of our time seems to be whether European man, the pinnacle of evolution, will destroy through the unnatural notions which are the modern products of his intellect what his inherited instincts have striven through these eons of time to preserve.[22]

In this, they are more true and faithful to the structure of ideas they believe in, than are the sociobiologists. For the latter tend to hide the politics of their theory behind a fact/value distinction.

So what would the living out of sociobiology be like? I don't only mean that it would require a reinforcement of male-female boundaries, for example, or of national xenophobic boundaries (though it would mean these). These are the *particular* expressions of its ideological life; the expression of the particular units that were taken up and scientized by it. The other portion of the specific ideological frame was the mechanism of genetic selfishness, and the proposal for stable patterns of interaction between them in an ESS. These were powered by "genes with pocket calculators", measuring their maximum chances.

It doesn't seem ludicrous to me to suggest that the implicit politics of this (which has grown by scientists' "innocently" reflecting into science the social world in which they live) is a proposal that we turn ourselves into commodities, calculating our selfish chances, but accepting that there is such a thing as a market (an ESS) that balances our individual strategies. But we must also accept that there is a State which has the job of regulating imbalances, "artificial" disorders, etc. In short, what worries me about sociobiology is its potential invitation to a unity of theory and practice in the latest forms of capitalist political economy.

It is a union of theory and practice in that the very things that the theory is about require, for consistency, their living out. That would involve refashioning human nature, remaking ourselves, along the lines required by the theory. As one popularizing article put it, "Perhaps there is a little bit of Maggie Thatcher in every one of us". Those who might resist can be adjudged "do-gooders", "intellectuals", "artificial";[23] and it will presumably be the task of the State to handle their miscreance.

Adopting a sentence from Hannah Arendt which I admire for its deep wisdom, "the trouble with modern reductionist theories is not that they are false, but that they might become true". We thus end with an irony that Darwinism thus conceived, with its mechanism in general of the non-rational causes of behaviour, becomes a potential mechanism itself for the "rational" transformation of humanity into its image. It is rational, at least, in the sense that its success depends in part on its acceptance as *true*, and the embodiment of that belief in a political praxis. Sociobiology would then be in theory and in practice the market come to life: reified forms taking dominance over people, and taking on the appearance of life – all, indeed, of the elements of an alienated life that Marx depicted. Reductionism is thus a political demand.

Notes and references

1. S. Rose, "Review of Chase: 'The Legacy of Malthus'," in *Race and Class*, no. 20, 1979, p. 243.

2. E. Fromm, *The Anatomy of Human Destructiveness*, Harmondsworth 1977, p. 22.

3. D. Bloor, *Knowledge and Social Imagery*, London 1977, pp. 5-6.

4. W. Trotter, *Instincts of the Herd in Peace and War*, London 1919, pp. 43-4.

5. R. Trivers, "The Evolution of Reciprocal Altruism", in T. H. Clutton-Brock and P. Harvey (eds.), *Readings in Sociobiology*, London 1979.

6. ibid., pp. 190-1.

7. R. Dawkins, *The Selfish Gene*, Oxford 1976, p. 4.

8. ibid., p. 35.

9. D. Barash, *Sociobiology and Behaviour*, London 1979, pp. 48-9.

10. N. Tinbergen, "On War and Peace in Animals and Man", in H. Friedrich (ed.), *Man and Animal*, London 1968, p. 122.

11. Consider the following quotation from the British *Daily Telegraph*, by a staff writer describing his reactions to an act of vandalism by "hooligans": "The mind-splitting question I was left with was how children who had obviously been so well cared for physically and materially could have been so disgracefully deprived of their right to restraint, discipline, and moral teaching. Those responsible for these offences against youth, even more so than the pederasts, the pornographers and the other exploiters, are the real ones for whom the millstones and the depths of the sea should be waiting" (R. H. Steed, "Who Trains Soccer Gangs?", 5 September 1977).

12. J. Quadagno, "Paradigms in Evolutionary Theory," in *American Sociological Review*, no. 44(1), 1979, pp. 100-109.

13. It is possible to accept such evidence without seeing it as evidence for the sociobiological thesis. Eric Jantsch can thus agree with Edward Wilson that beyond the social insects there has been a tendency for species to develop individuation. Whereas the sociobiologists seek an immediate genetic explanation, Jantsch derives a phyletic explanation from Walter Freeman's evidence. Freeman found in experiments with cats and rabbits sudden short bursts of EEG activity when they had to notice and take stock of unfamiliar objects. Jantsch, quite reasonably, takes this to be evidence of a form of mental modelling, formation of individuated perceptions which would therefore be associated with an increased individualism in behaviour. See E. Jantsch, "Self-realisation through self-transcendence", in E. Jantsch and C. H. Waddington (eds.), *Evolution and Consciousness*, London 1976, pp. 53-4.

14. D. Barash, op. cit.

15. S. Rose, *The Conscious Brain*, Harmondsworth 1976, p. 28.

16. G. Cutten, *Mind: Its Origin and Goal*, Boston 1925, pp. 13-14.

17. W. McDougall, *The Group Mind: A Sketch of the Principles of Collective Psychology with Some Attempt to Apply Them to the Interpretation of National Life and Character*, Cambridge 1920.

18. It will be objected that even reductionist biologists accept the reality and significance of learning. That is true. But if we look at their typical view of learning, they stress its continuity with processes of genetic evolution. Thus we have genetic mutation described as equivalent to "trial and error learning", evolution described as a "genuine cognitive process"; in reverse direction, language and culture have been described in terms of "survival of ideas", the development of scientific ideas as matching genetic mutation with its "storing up of advantages", and so on. The whole tendency to describe consciousness in ways analogous with *processes* of evolution seems to me to be very suspicious. Once

again, in the process the specificity of consciousness, culture and learning are lost.

19. K. Marx, *Early Writings*, Harmondsworth 1975, p. 328.

20. On this, see Waddington's conclusion: "A surprisingly large amount of the environment which exerts natural selection on an animal is the more or less direct result of the animal's own behaviour" (C. H. Waddington, "Evolution in the sub-human world", in Jantsch and Waddington, op. cit., p. 13).

21. Consider the following examples of an opposition between "natural" and "artificial": "In many carnivorous animals . . . a mechanism exists that causes them to prefer food with a minimum content of fibre and a maximum of sugar, fat and starch. In the 'normal' conditions of wild life, this phylogenetically adapted releasing mechanism is of obvious survival value, but in civilized man it gives rise to a search for supernormal objects, the addiction to which actually amounts to a vice detrimental to health (e.g. white bread, chocolate, etc.)" (K. Lorenz, *Evolution and Modification of Behaviour*, Chicago 1965, p. 17). And: "Earlier in this chapter we considered why sugar is sweet. Our adaptive fondness for sugars has actually become a major nuisance now that cultural evolution has discovered it can pander to this biologically-evolved trait. Through culture we provide ourselves with the essentially non-nutritive cakes, cookies, chocolates and assorted affluvia of the sweet-tooth industry. We know it's not good for us, but we are still biological creatures and can't easily resist our fondness for sugars" (D. Barash, op. cit., p. 322).

22. R. Verrall, "Sociobiology – the instincts in our genes", in *Spearhead*, May 1979.

23. I borrow the terms from British discussions of immigration, where such monsters were counterposed to the "natural feelings" of ordinary people about immigrants. See my article, "Racism: the New Inheritors", in *Radical Philosophy*, no. 21, 1979, pp. 2-17, in which I traced the correlation with ethology and sociobiology. See also my *The New Racism: Conservatism and the Ideology of the Tribe*, London 1981.

2
Human Nature and
Social Explanation
Janna L. Thompson

Introduction: two forms of reductionism

We are biological organisms. We have bodies and nervous systems. We are also people who live in a social world where we are teachers, workers, wives, etc. We have a genetic constitution which we have inherited from our ancestors and may pass on to descendants. From past generations we have also inherited social institutions and practices. We are in some sense the products of biological evolution – this we have in common with other creatures of the world. But we also have a language and a culture which undergoes historical change – in this respect we seem to differ from all other creatures. What is the relationship between our existence as biological and social beings?

Attempts to answer this question usually take one reductionist path or another. Facts of one kind are ignored, ruled out, or are explained in terms of facts of the other kind.

Social roles and institutions are the result of the actions and beliefs of individuals. From this uncontroversial starting point generations of philosophers, psychologists and others, from John Stuart Mill to B. F. Skinner, have leaped to the reductionist conclusion that statements about social facts must be derivable from statements about the psychology of individuals – statements which must not themselves presuppose any social categories or descriptions. Under the influence of positivism, it is a short step from a position of "methodological individualism" to biophysical reductionism, the idea that psychological statements in their turn need to be reduced to statements about physiology and biology. To those reared on the positivist view of explanation, biophysical reductionism is so much a matter of common sense that it doesn't need to be argued for. For example Edward Wilson in *Sociobiology* proceeds as if the only thing he needed to establish is that biological theory will play a larger role in the reductionist strategy than is usually assumed. From this point of view behaviourism – another version of reductionism – is sociobiology's only serious opponent; and establishing the need for sociobiological explanations of social behaviour means establishing that behaviourist theory is not adequate for reductionist purposes.

In opposition to Mill's methodological individualism, and to Herbert Spencer's attempt to locate the foundations of the social world in biology, Emile Durkheim aimed to establish the study of human society as an independent scientific discipline with an irreducible subject matter.[1] Durkheim, the structuralists and the structural-functionalists who followed him, insist that social facts irreducibly exist, and that they should be explained only in terms of other social facts. Biology and psychology are ruled out of bounds when they try to invade the realm of social theory. To Marshall Sahlins, who defends anthropology against the sociobiologists in *Use and Abuse of Biology*, this position is equally one of common sense.

The belief that social phenomena can only be explained in terms of theories about social facts is a reductionist position – what I will call sociological reductionism. These reductionists are not, of course, denying that biological and psychological facts exist – in their own proper sphere (although they do argue that many psychological facts turn out, under closer examination, to be social facts). But in respect to social theory, bio physical reductionism and sociological reductionism are parallel theses: one insists that social phenomena can only be accounted for in terms of biological, psychological or physiological theories; the other that social phenomena can only be explained by sociological theories.

In this paper I will argue that both forms of reductionism are inadequate and that an attempt to relate the biological and the social must be critical of both.

Sociobiology and reductionism

Sociobiology is not so much a discipline as an undisciplined collection of theses and models for relating the biological and the social. These theses do not say the same thing, and they are not all compatible. This is one of the reasons why sociobiologists are so difficult to attack. When one of their strong positions is no longer defensible, they retreat to a weaker position from which they complain of being misunderstood. When the coast is clear, they reassume their former standpoint. The confusions inherent in sociobiology are particularly evident in Wilson's writings, in both *Sociobiology* and *On Human Nature*.

Sometimes Wilson is as committed to biophysical reductionism as anyone can be:

> The transition from purely phenomenological to fundamental theory in sociobiology must await a full, neuronal explanation of the human brain ... Evolutionary sociobiology will ... attempt to reconstruct the history of the machinery and to identify the adaptive significance of each of its functions.[2]

There doesn't seem to be much of a place for social facts in this scheme of things. Social theory as we know it is not to be merely subsumed; it is to be wiped off the scientific map. Social facts are, at most, epiphenomenal; the real explanation of "social" events is to be provided by physiology and biology. At worst, social theories will prove to have the same scientific stature as statements about witchcraft. At best, they hint at where the true explanation lies.

In *On Human Nature*, Wilson's position generally appears to be more moderate. Social theory is to be reduced to physiology and biology in the same sense that chemistry is reduced to physics: its categories and theories are to be derived from the categories and theories of the more basic sciences.[3] Social theory cannot lose its legitimacy or independence as a science as the result of such a reduction; in fact the possibility of such a reduction presupposes the truth of at least some of the explanations of social theory.

Sociobiologists, nevertheless, have an ambivalent attitude toward social theory – an ambivalence that is, perhaps, the result of a vacillation between these two forms of reductionism. Wilson talks of the "blending of biology and the social sciences" which will result in the joining of the "two cultures" of Western intellectual life.[4] On the other hand, Wilson and other sociobiologists are only occasionally interested in the theories of social sciences – and then, only when they want to try to show that these theories are wrong. In practice, sociobiologists seem to be in the business of replacing social theory rather than complementing it or explaining it on a deeper level. Sociobiology is supposed to provide an explanation for the irrational, for the primitive emotions governed by the hypothalamus which are not easily made subject to reason or the pressures of social conventions.[5] Sociobiology is supposed to step in when anthropologists and sociologists can't provide adequate explanations for social phenomena (like incest taboos).[6]

On the other hand, sociobiologists sometimes claim to provide the key to a deeper rationality behind human actions – as opposed to the reasons of people themselves, or of social theorists. This deeper rationality is, of course, the rationality of gene preservation. Sometimes sociobiologists' claims on behalf of this rationality are ambitious: it becomes the real reason behind most of our social actions and institutions from sex role behaviour to religious mysticism. [7]

From this interesting but improbable position, sociobiologists retreat to a number of less interesting but more defensible views. Sometimes the biological programming is generalized by being made less precise: thus we no longer have an urge to be altruistic according to the degree to which other people are biologically related to us, but a more generalized urge to be nice to people who are close to us. Sometimes the "urge" itself is diluted: it's always there but can be overridden by contingencies. Usually, to be on the safe side, sociobiologists both generalize and dilute these innate inclinations or drives.

Our biology, then, is supposed to endow us with a predisposition to do something or other. But this can be understood in different ways. Sometimes it is meant to be a statement about the urges that dwell in every human hypothalamus. Sometimes when sociobiologists remember that they're supposed to be basing their theories on population genetics, it's a statement about behaviour resulting from gene frequencies in a population. Sometimes the predispositions are weak and allow for frequent deviations; sometimes their strength is supposed to be overwhelming. And sometimes they dwindle to a repertoire of emotions and capacities – to what even a dogmatic environmentalist would probably be prepared to accept. [8]

The explanatory efforts of the sociobiologists can be sorted into two categories. The first are those which attempt to explain some aspects of human social behaviour (whether a lot or just a little) as the direct result of the operation of biophysical causes. The second are the efforts of those who insist that biology does affect human social behaviour in more ways than are presently admitted (and that it is possible to determine what this biological contribution is), but believe that biological influences are inevitably filtered by the socialization process and a person's perception of herself in her social world; no aspect of social behaviour can therefore be attributed directly to a biological cause. The first are reductive explanations, whether they are put forward in the framework of a general reductionist thesis or not. But the second need not be reductionist at all. The sociobiologists are, of course, renowned for explanations of the first kind. But it is possible to find examples of the second in their work. And perhaps because of this, sociobiology has received sympathy from some unexpected quarters.

Sociobiologists who attempt to give reductive explanations of social phenomena do not have to deny the existence of social facts or the validity of all sociobiological and anthropological explanations (although as philosophical reductionists, they may believe that science will some day be well rid of them). What they do claim is that some aspects of human social behaviour, and thus of human society, can be explained as the direct causal consequence of genetic "programming", and ultimately of a theory of natural selection.

The operation of the causal mechanism takes place, so to speak, behind the back of consciousness and culture. That is, in so far as consciousness, belief or cultural influences enter the picture, they do so as effects, as the consequences of genetic causes. Sometimes sociobiologists admit that the way an agent understands herself, her acceptance and adherence to social norms, can modify or even contain the operation of this causal mechanism (as Lorenz, for example, suggests in *On Aggression* that our natural aggressive impulses can be checked or deflected by morality). But this admission is a dangerous one. For to the extent that consciousness is able to alter the way a biological mechanism operates, the resulting behaviour cannot be said to be caused directly by the mechanism.

In so far as these reductive explanations allow for social change at all, they admit it only as the result of technological manipulation. Change has to be a matter of manipulating people or their environments in order to modify the cause or the way it operates.[9] Either people have to be put in an environment which doesn't trigger off the mechanism (as is done, for example, to prevent the genetically caused disease phenylketonuria) or the biological forces involved have to be met with an equal and opposite force. Like Hobbes, sociobiologists generally assume that going against the grain of human nature requires a considerable amount of authoritarianism. The regulations required to eliminate all sexual differences in behaviour, Wilson argues, "would certainly place some personal freedoms in jeopardy."[10]

If the purpose of social theory is to make the social world intelligible to social agents so that they can act together to change this world, then even if sociobiology were able to provide correct predictions about human behaviour, it would have only the kind of relevance to social theory that medicine and physiology have now. It would give people information about themselves which they would have to take into account. The claim that sociobiology *is* social theory, or an important part of social theory, is a political claim, and therefore is rightly criticized on political grounds.

However, the reductive explanations of the sociobiologists are not successful, by any scientific measure of success. Their attempts to account for social behaviour as the causal result of a genetically derived mechanism do not explain what they are meant to explain; and when they are altered to make them more plausible, they cease to be reductive explanations. This criticism can only be advanced by examining cases, something which I cannot do exhaustively here.

It is easy work to point out the inadequacies in the ambitious reductivist efforts of the sociobiologists – their attempts to explain altruism, adultery, the generation gap, etc. as behavioural consequences of a mechanism which "calculates" genetic costs and benefits. These attempts involve redefining terms like "altruism" so that they become amenable to the sort of explanation which sociobiologists want to give; and then require a suspension of disbelief of heroic proportions. For it is necessary to keep up the pretence that "altruism", as it is now used, still means what it usually means; it is necessary to ignore behaviour which does not conform to the new definition.

When sociobiologists are forced to acknowledge the implausibility of biological determinism, they begin their retreat into a more easily defended position, making as few concessions as possible. Wilson admits, for example, that not all aspects of religious behaviour can be biologically explained, but in *On Human Nature* he ambitiously sets out to explain the major features of religious practices as the causal result of biological programming. His idea is that these practices are the end result of characteristics of individuals – their willingness to conform, to sacrifice themselves, to be indoctrinated, to make myths – characteristics which in turn are the result of "neurologically based learning rules that evolved through the selection of clans competing against one another."[11] Conformity, etc. is supposed to increase the reproductive potential of these kinship groups.

There is, of course, an immense gap between religion on one hand and self-sacrifice, conformity etc. on the other. Wilson tries to narrow it by widening the term "religion" to include marxism, nationalism, even scientific materialism – presumably the term must take in any ideal which can excite devotion and self-sacrifice. One problem among many is that ideals are such that people can share them with others who are not their kin. (It would seem that our genes could have found a less dicey way of perpetuating themselves.) It is probably not surprising, then, that Wilson seems to slide from a kinship selection thesis to a group selection thesis as "clan" comes to have a wider and

wider enrolment. From this point, it is only a relatively small conceptual leap to Wilson's final functionalist characterization of religion:

> In the midst of the chaotic and potentially disorienting experiences each
> person undergoes daily, religion classifies him, provides him with
> unquestioned membership in a group claiming great powers, and by this
> means gives him a driving purpose in life compatible with self-interest.[12]

But this is a sociological, not a biological, account of religious belief. It
has to do with the way people perceive themselves and their needs in
their social worlds, and the way they are acted on by social pressures.
The genetic mechanisms have vanished without any obvious trace.

The same metamorphosis from the biological to the sociological
occurs again and again in sociobiology, even when less ambitious
explanatory labours are being attempted, for example incest prohibi-
tions.

There is no room or reason to labour the point with further examples: it seems likely that any attempt to explain an aspect of social
behaviour as the result of a cause which operates independently of
beliefs and cultural context is doomed to run into the problems I have
been discussing. However, this criticism of reductive explanations
does not preclude the possibility that biology and biological explanations may play an important role in social theory.

On being predisposed

Not all of sociobiology's fellow travellers are prepared to accept its
reductive explanations of social behaviour. But many are supporters
of the sociobiological enterprise because they believe that biological
facts and theories could play a greater role in social theory than they
do, and thus they look forward to a more intimate relationship between biology and social theory than most social theorists are now prepared to contemplate.

That biological phenomena can have an effect on social behaviour is
something difficult to deny. Humans as biological beings have certain
biological needs which societies have to accommodate in one way or
another. For instance, people need to consume food at fairly regular
intervals. The physiological basis of hunger can be identified and
measured. Biological data and theories can tell us what foods people
can eat: which are poisonous, which nutritious, and why. On the
other hand, eating is a social phenomenon invested with social mean-

ing. Exactly what and when people eat, how they eat it, what social functions eating can fulfil is, of course, a matter for social theory to describe and explain.

What the non-reductionist friends of sociobiology seem to be saying is that there are other less obvious biological needs, propensities or states, which also influence social behaviour. They want to say that our biology provides us with more than a stock of basic emotions and feelings and a capacity to reason, but also certain urges, desires, ways of thinking. How and when these urges are expressed, how they are perceived, depends on social and personal factors. The urges are always filtered through the conventions, practices and understandings of a social world.

The friends of sociobiology believe that it is plausible to suppose that we have biological predispositions and possible to find out what they are. I am going to look at two attempts to make these claims plausible.

A person has a predisposition to do something or feel something if in some specifiable circumstances she is likely to do that thing or have that feeling. Predispositions can't be postulated into existence. Those who claim that some propensities are biologically based must have some way of distinguishing the influence of the environment from the influence of the genes. This isn't easy, for propensities are by nature elusive. Wilson argues, on the basis of a comparison between apes and humans, that humans have a slight propensity to be polygamous.[13] This predisposition (if it exists) doesn't of course explain why a particular society practises polygamy. This, he admits, will have a social explanation. But, further, the existence of such a predisposition is not only perfectly compatible with the existence of non-polygamous societies, it is also compatible with there being no polygamous societies at all. Further, the prevalence of polygamous societies is compatible with the possibility that there is no biological predisposition for polygamy. Polygamy may be a response to a common environmental and social problem.

The difficulty is further compounded by the fact that sociobiologists tend to be very unspecific about what our predispositions are and under what circumstances they are supposed to be manifested. Sociobiologists, as we have seen, have a predisposition to dilute and generalize their predispositions until they threaten to turn into mere capacities. Nevertheless, sociobiologists and others have made some suggestions about how to distil the biologically based elements out of the complexity of social behaviour.

One suggestion is this: that biological propensities are likely to be at work when behaviour is persistently irrational or when a social practice is dysfunctional. For instance, it is fairly common to suggest that warfare, since it is irrational and socially destructive, must be to some extent the consequence of biological propensities.[14] Or that phobias – e.g. the fear of snakes – being irrational, are the result or partial result of biological propensities. (The idea is that a propensity to be fearful of snakes is useful for members of a hunter-gatherer society.)

The problem is that what we regard as irrational or dysfunctional, and thus what calls for a biological explanation, depends on what theories about society and rationality we happen to accept.

In any case, do the irrational or the dysfunctional need a biological explanation? The idea that dysfunctional features of society are somehow beyond the ken of social theory may simply be the result of a determination to believe that all social practices and institutions must be functional parts of a harmonious whole. There are social theories, like marxism, which can accommodate social irrationality without bringing in biology.

There are also alternative theories in the business of explaining irrationality in human behaviour. Fear of snakes, for instance, is accounted for in Freudian theory by connecting it with sexual fears and phantasies.

Mary Midgley in *Beast and Man* embarks on another route to the biology behind social behaviour. Human behaviour, she argues, can't be regarded as the result of calculation or conditioning. Behaviour is motivated; beneath it lurk wants, desires, urges, feelings. It is these motivations that must be examined in an effort to uncover the biological substratum.

Midgley admits that wants arise in a social framework. Nevertheless, she seems to believe that sophisticated social wants must be derived from our primitive, earthy wants. Sometimes she seems to think that determining what these wants are is rather like a process of skimming off cream. You remove the culturally conditioned factors and what is left is the biological:

> Moreover, our basic repertoire of wants is given. We are not free to create or annihilate wants, either by private invention or by cultures. Inventions and cultures group, reflect, guide, channel, and develop wants; they do not actually produce them. Thus if twentieth-century people want supersonic planes, they do so because of wants that they have in common with Eskimos and Bushmen. They want to move fast, to do their business quickly, to be honoured, feared, and admired, to solve

puzzles, and to have something bright and shiny. We are innately "programmed" to want and like such things.[15]

The procedure here is similar to what goes on in a lot of sociobiology. The idea is that you can get to the biological heart of the matter by making a description of an action or motivation more vague and general. Even if this were all that was required, Midgley's end product would still be unsatisfactory. Not all groups of people want to move fast and do their business quickly. To understand why some people want the Concorde we have to understand why speed and efficiency is regarded as so important by some people in our society. And the reason why some people find large pieces of complicated technology exciting is probably best explained by reference to ideas about technology that have developed in our society. Nothing much is accomplished, as far as explanation is concerned, by bringing in a supposed natural attachment to bright, shiny objects. People everywhere probably want to be admired, honoured, if not necessarily feared; everyone usually wants to be successful in their undertakings. But there is no guarantee, even when we have got this far, that we are down to biological basics, or that there are any biological basics to come to beyond the capacity to love and fear, to think and feel. There may be biologically based urges lurking behind our various wants, but they can't simply be read off from observations of behaviour, or created by a general description of a motivation. The only way of approaching them is by way of an examination and critique of theories of behaviour and motivation. The only way to establish their existence is to show that theories which try to get on without them are unsatisfactory.

Elsewhere Midgely recognizes this. She discusses the case of a man who buys a house with an acre of land because he insists that he hates to be overlooked by strangers. She then attempts to show that this motive, and the desire for privacy in general, cannot be explained adequately by Freud or Marx, and thus that the sociobiological explanation – that humans, like other creatures, have a biologically based predisposition to avoid being stared at – is a plausible alternative.[16] While her dismissal of Freud and her discussion of Marx may be less than convincing, this indirect approach to the biology behind behaviour seems to be the only one that has any chance of success.

The non-reductionist friends of sociobiology should face up to the possibility that in their search for the biological substratum of human behaviour they may well be on the trail of a chimera. For it may be that the biological and the social are so inextricably welded together in human thought and action that they cannot be prised apart by any theoretical tools. Sociobiologists, whether reductionist or not, found their investigation on the assumption that it is possible to separate the contribution of the genes from the contribution of learning. To question this, as many of this conference's participants have done, is to question the whole sociobiological enterprise. But it is not to reject the possibility of a theory of human nature, or to deny that biological facts and theories may enter into or influence social theory.

Sociological reductionism

For Durkheim, and later for Sahlins, the argument against biological or psychological reductionism seems at first to be an empirical one. Sociological phenomena are simply too complex to be explained as the consequences of biological or psychological phenomena:

> What an abyss, for example, between the sentiments man experiences in the face of forces superior to his own and the present religious institution with its beliefs, its numerous and complicated practices, its material and moral organization![17]

Similarly Sahlins argues in *Use and Abuse of Biology* that the nature of kinship systems, the rules and conventions concerning them, cannot be understood as the consequences of biological theories of kinship selection.[18] I have been advancing similar arguments in the earlier parts of this paper.

However Durkheim's insistence that only social facts can explain social facts, and Sahlins's opposition to sociobiology, sometimes seem to depend on a stronger position – a position which rules out a priori any kind of biological or psychological entry into social theory.

"Social facts", says Durkheim, "do not differ from psychological facts in quality only: they have a different substratum; they evolve in a different milieu; and they depend on different conditions."[19] By constituting a social fact as thing, Durkheim sometimes seems to believe that he has identified a kind of Hegelian subject, in respect to which human individuals and their feelings and beliefs are predicates.

> To understand the way in which a society thinks of itself and of its environment one must consider the nature of the society and not that of the individuals. Even the symbols which express these conceptions change according to the type of society.[20]

Here then is a category of facts with very distinctive characteristics: it consists of ways of acting, thinking, and feeling, external to the individual, and endowed with a power of coercion, by reason of which they control him. [21]

The conceptual separation between social facts and individual qualities and desires seems as total as the Cartesian distinction between mind and body, and is put forward for the same reasons. The terms and theories used to describe and explain the social "whole" must be different from the descriptions and explanations of the individuals and their psyches which makes it up. Social phenomena "cannot be reduced to their elements without contradiction in terms, since, by definition, they presuppose something different from the properties of these elements."[22]

Sahlins's conception of society and social explanation is not the same as Durkheim's, but nevertheless he also suggests that there is an insurmountable conceptual boundary between the sociological and the biological:

Culture is the essential condition of this freedom of the human order from emotional or motivational necessity. Men interact in the terms of a system of meanings, attributed to persons and the objects of their existence, but precisely as these attributes are symbolic, they cannot be discovered in the intrinsic properties of the things to which they refer. [23]

Later Sahlins draws out the implications of this conception of culture:

The reason why human social behaviour is not organized by the individual maximization of genetic interest is that human beings are not socially defined by their organic qualities but in terms of symbolic attributes; and a symbol is precisely a meaningful value . . . which cannot be determined by the physical properties of that to which it refers. [24]

The meanings of the terms used in social life and thus the social phenomena which they denote are constituted by their role in this system of social meanings.[25] And this system of social meanings is impervious to biological terms and explanations. It is true that biology, in a sense, makes culture possible (as brains make mental activity possible and provides phenomena which a culture invests with symbolic meaning. But because cultural events are symbolic events, "a radical discontinuity is introduced between culture and nature."[26]

Some marxists insist for similar reasons that views about human nature must be purged from a scientific social theory, whether these views come from biological theory or an ethically inspired ideal. In *For Marx* Louis Althusser takes this position against humanist marx-

ists and the early Marx, who regard capitalism and other forms of
social exploitation as an alienation of the human essence, and
revolutionary change as the liberation of this essence. It is not simply
the prescriptive use of "human nature" in the early Marx which
Althusser regards as unscientific:

> If the essence of man is to be a universal attribute, it is essential that con-
> crete subjects exist as absolute givens; this implies an empiricism of the
> subject. If these empirical individuals are to be men, it is essential that
> each carries in himself the whole human essence, if not in fact, at least in
> principle; this implies an idealism of the essence.[27]

A social theory which accepts human individuals as given is being
"empiricist" in this derogatory sense. The real "subjects" of history –
and thus of social theory – are "given human societies".[28] The mature
Marx "replaced the old couple individuals/human essence in the
theory of history by new concepts (forces of production, relations of
production, etc.)".[29] "Humanity" is not a category in social theory.
Human beings and human needs have no place there. Humans enter
social theory as proletarians, capitalists, or whatever; needs enter as
social needs.

There are, then, two closely related arguments for a radical discontinuity between social theory and views about human nature –
wherever these come from. They are, in fact, basically the same
argument expressed in different ways.

(1) In social theory events and actions are described and explained
in the context of social roles, conventions, rules and practices of the
society in which this event or action takes place. The identity as well as
the significance of social events are determined by their role in this
social web of meaning. Therefore the theories of biology, or psychology, or any theory about human nature, cannot be about the same
things that social theory is about – even when the same words are
used. The consumption of edibles is not the same as eating. The
hunger satisfied with a knife and fork is different from the hunger
satisfied by raw meat.[30]

(2) The subjects of social theory are social structures or social
systems, and understanding social events or institutions means
describing and explaining them so that they can be seen as interrelated
parts of such wholes. Anything that can't be so described or explained
is not part of social theory (though it may be treated by other sciences
or by a theory of ideology).

I have suggested that this way of defending the boundary between social theory and theories about human nature is similar to the way in which Cartesians defend mind from the pollution of body. In fact sociological reductionism has within it some of the same problems that are inherent in dualism. For sociological reductionists admit (as dualists do about the relation between mind and body) that there is some interaction between the biological and the psychological on the one hand, and the social on the other. But they leave us with no way of understanding this relationship, and no way of theorizing about it. For social theory cannot do it; and neither can any theory about "humanity". The radical discontinuity between the two does not allow for the possibility of a bridge.

Sociological reductionism also leads to relativism of various kinds. It tends toward cultural relativism – to the view that different social structures, or the "same" social structure before and after a revolutionary break, are incommensurable.[31] For the components of a whole or a web of meaning cannot possibly be the same as the components of another whole or another web of meaning, since their identity is determined by their being an integrated part of a particular social whole. The problem of cultural relativism is the more immediate, for it challenges the very possibility of social theory, as something that goes beyond a descriptive and theoretical account of a particular society.

General theories and descriptions of social structure pose several problems for the reductionist. First of all the sociological reductionist does not escape from relativism by these means. The problem reappears at another level. Social structures, however they are defined, remain incommensurable. If capitalism and feudalism are two different social types, then there seems to be no way to compare the phenomena in one with the phenomena in the other. What we call a "class" in one bears no relationship to "classes" in the other. How then is a theory of the historical development of societies possible?

Further, the sociological reductionist allows us no way of understanding the relationship between the social type and the social particular. The type is itself a whole with its own web of relationships. How then can the terms and theories used to account for it be relevant to a particular social structure which, after all, is a different whole, with a different web of relationships?

I don't have the time, energy or capacity to untangle these problems and deal with them all satisfactorily. But I want to suggest that one thing that prevents cultural relativism from being an insuperable problem, one reason why social theory is possible, is that we approach social theory with some notion of what human beings are and what needs they have. This notion may be extremely rough and vague, but nevertheless it gives us a starting point for considering what sorts of problems humans as social beings must solve. If this is so, humanity and human needs must in one way or another be the concern of social theory.

Human nature and marxist social theory

> The premises from which we begin are not arbitrary ones, not dogmas, but real premises from which abstraction can only be made in the imagination. They are the real individuals, their activities and the material conditions under which they live, both those which they find already existing and those produced by their activity. These premises can thus be verified in a purely empirical way. The first premise of all human history is, of course, the existence of living human individuals. Thus the first fact to be established is the physical organization of these individuals and their consequent relation to the rest of nature.[32]

The German Ideology, like much of Marx and Engels's work, is an attempt to comprehend human society – its evolution and its potential for change. Marx in his later works concentrates largely on analysing a particular social structure – capitalism – but his social theory, early and late, depends on a general view of the relationship between human needs, society and the natural world – what is sometimes called the dialectics of humanity and nature.

The beginning point of Marx and Engels's analysis of the social world are the real individual human beings who face a real natural world. They insist, against the German idealist philosophers, that neither of these two realities can be dissolved away into an ideal entity. The form that the social world takes is conditioned, but not determined, by these "givens". It is a mistake to explain features of human society using the terms and theories of physics, biology or psychology. But it is also a mistake to believe that our biological or psychological being can be reduced to a system of social relationships.

Marx and Engels's starting point influences the way in which they analyse societies – i.e. what they focus on, what they believe is important to explain. Their starting point is what makes it possible for them to have a general theory about society and social evolution.

For they begin with the fact that humans have certain important natural needs and a certain kind of physical being. We need food, shelter, etc. And although this seems to be a trivial thing to say, it has a central importance for Marx and Engels. What exactly we want to eat or wear, and how we go about getting these things, is going to depend on the form of society we are in and the natural environment we face. But the fact that we know that people need food, shelter, etc. provides us with a way of understanding some of the actions of people in any given society (however strange they may otherwise be) and a way of comparing features of one society with those of another (no matter how they may otherwise differ). Whatever else people do, however they conceive themselves and their activities, they are going to be preoccupied to a greater or lesser extent with satisfying these elementary needs. And further, since these needs are important, many of the other activities of people and their ways of looking at themselves and the world are going to be affected by the means they use to satisfy them. This is one of the reasons why Marx and Engels believe that we can understand societies and social evolution by focusing on production.

Marx and Engels mention only elementary and obvious needs in this section of *The German Ideology*. It is possible that humans have less obvious natural needs, desires, propensities which also condition and are conditioned by their social existence. While Marx and Engels's view of social theory is clearly incompatible with biophysical reductionism, it does not seem to me that their dialectical account of social evolution is incompatible with the existence of less obvious natural propensities.

One way of understanding Marx and Engels's insistence that social being determines consciousness, does not require that mention of human beings, or human needs, abilities, propensities, be eliminated from social theory. The existence and development of such needs, propensities etc. may indeed depend on the existence of social relationships – the qualities and abilities that are specifically human may be inconceivable apart from social relationships of some form. But this doesn't preclude the possibility that there are universals of human social behaviour or that there may be such things as natural needs, abilities and propensities. This understanding of Marx and Engels would not, for example, rule out Chomsky's theory of a universal deep structure or Jurgen Habermas's theory of the universals of human communication, or for that matter, other non-reductive theories about "human nature".

This interpretation of Marx and Engels's dialectics is less prone to being knocked down by possible developments in the theory of language, etc. I want to argue that it is also the only interpretation compatible with marxism considered as a theory of practice – a theory which is supposed to bring about social change by becoming understood, accepted and used by the working class and its allies.

Most modern marxists agree that marxism should not be understood as a theory about the laws of motion of capitalist society; that it is not a theory intended to chart the inevitable downfall of capitalism as a consequence of the contradiction between the forces and relations of production. If this were all there was to it, then class consciousness would simply be a product of this contradiction and socialist society the inevitable outcome of it. But if we don't understand marxism as a mechanical theory of revolution then class consciousness must have a larger and more creative role to play; for whether a revolution takes place at all, what form it takes and what a future society will be like depends to a considerable extent on the nature of class consciousness. Since this class consciousness isn't just a product of social relations, then what values people have, what ideas they have about what a society should be like and what sort of life people should lead are important. A social theory which is a theory of practice (as opposed to those sociological theories which are thought to be value free) therefore needs a system of values which is not only connected with the needs and values people now have, but which can provide a justification for adopting values and ways of thinking which many people may not yet have. A social theory which is a critical theory – a theory which evaluates as well as describes and explains – needs an ethical standpoint from which to make its evaluations and a system of values which is not reducible to what is good for a particular class at a particular historical period. Without such a system the notion of historical progress doesn't make any sense.

From the *1844 Manuscripts* to the *Critique of the Gotha Programme* (1875) Marx bases his evaluations on a view about human needs, abilities and potentialities. For this reason his appeal to human needs, and to an ethics based on these needs, cannot be dismissed as a Hegelian residue which Marx grows out of after 1844.

Althusser, who does so regard it (and who is not a mechanistic marxist), faces a serious problem: if human needs and capacities are eliminated from marxist social theory, then so is the system of values that goes with it. There is now no room for evaluation except as ideology.[33] Althusser recognizes that this makes ideology indispensible – a necessary adjunct of revolutionary consciousness and not something that can be overcome by it. But this is not a satisfactory position, for now it seems that the marxist social theorist must on the one hand deny values any objectivity, and at the same time has to pretend that they are objective when she is talking to working-class people or engaging in social practice herself.

Should we return, then, to the standpoint of the early Marx and his theory of species-being? The problem is that Althusser's suspicions about this notion are justified, and that further, it seems clear that Marx and Engels, when they wrote *The German Ideology*, weren't happy about it either. What Marx does in his essay on alienation in the *1844 Manuscripts* is to take Feuerbach's conception of humanity more or less as read, and apply it to the critique of alienation in capitalist production. Humans are universal beings and "free, conscious activity is man's species character".[34] But this activity is thwarted in a society where relationships are exploitative. In this characterization of human essence facts and values are welded together, and Marx nowhere tries to prise them apart in order to discuss what their relationship should be. Nor does Marx attempt (as he does in *The German Ideology*) to relate the needs that arise from human abilities to the needs that arise in a particular society at a particular historical period. There is no doubt that marxists, following in the footsteps of Marx himself, rely on ideas about what motivates human beings and what potential people have for change. These ideas are rightly called myths, for like Marx's early concept of species being, they have no better backing than a philosophical tradition and wishful thinking. Rather than attempting to eradicate all views about humanity, as such, from social theory, we should be searching for a "scientific" theory of human nature, a theory which can be supported by evidence and argument.

In this search, sociobiology as we know it is not likely to be of much help. But this doesn't mean that a theory of human nature is impossible. Such a theory, presumably, would identify and investigate our specifically human capacities and motivations. Such traits are not likely to be simply biological, but rather the consequences of our biological endowment and the learning that takes place in any human environment. The theory would help us to explain historical change and predict the possibility of future changes. It could also serve as a basis – as factual theories do – for a view about what the goal of social action should be.

There are a number of approaches to this theory which could prove fruitful: the Hungarian and Yugoslav Marxists' search for a humanist marxism; Jurgen Habermas's attempt to ground the political aims of marxism on an analysis of the prerequisites of human communication;

Levi-Strauss's search for the human ways of thinking behind cultural variation; the neo-Freudian attempt to unite a historical approach with a view about the springs of human action. It is with such theories that we should begin a search for a scientific understanding of the relation between the biological and the cultural. The fact that it is not any of these, but the red herring, sociobiology, which has captured the imagination of the scientific community and the attention of the press, is an indication of the reductionist proclivities of our thinking and the reactionary climate of our times.

Notes and references

1. For Durkheim's discussion of sociological method see G. Catlin (ed.), *Rules of Sociological Method*, Chicago 1938.

2. E. O. Wilson, *Sociobiology: the New Synthesis*, Cambridge, Mass. 1975, p. 525.

3. E. O. Wilson, *On Human Nature*, Harvard 1978, pp. 6-10.

4. ibid., p. 10.

5. See Wilson on phobias in *On Human Nature*, p. 68ff.

6. ibid., pp. 36ff. and 68.

7. One of the most apparently uncompromising statements of this position is at the beginning of Richard Dawkins's *The Selfish Genes* (Oxford 1976): "We are survival machines – robot vehicles blindly programmed to preserve the selfish molecules known as genes". Nevertheless, Dawkins later admits that as creatures with consciousness and culture, we have escaped this biological determinism, and in the end isn't prepared to say how much, if any, of our behaviour is governed by the "selfish gene".

8. "The behavioural genes more probably influence the ranges of the form and intensity of emotional responses, the thresholds of arousals, the readiness to learn certain stimuli as opposed to others, and the pattern of sensitivity to additional environmental factors that point cultural evolution in one direction as opposed to another" (*On Human Nature*, p. 47).

9. Wilson suggests in the conclusion of *Sociobiology* – that sociobiology is a tool for social managers.

10. *On Human Nature*, p. 133.

11. ibid., p. 184.

12. ibid., p. 188.

13. ibid, p. 20.

14. Lorenz suggests this and so does W. H. Durham in a discussion of primitive warfare (see Michael Ruse, *Sociobiology: Sense or Nonsense?*, Dordrecht 1979, p. 172).

15. Mary Midgley, *Beast and Man*, New York, 1978, p. 183.

16. ibid., p. 6ff.

17. *Rules of Sociological Method*, p. 106.

18. M. Sahlins, *The Use and Abuse of Biology*, London 1977, ch. 1.

19. *Rules of Sociological Method*, p. xlix.

20. ibid., p. xlix.

21. ibid., p. 3.

22. ibid.

23. *Use and Abuse of Biology*, pp. 11-12.

24. ibid., p. 61.

25. ibid., p. 62.

26. ibid., p. 12.

27. L. Althusser, *For Marx*, London 1969, p. 231.

28. ibid., p. 231.

29. ibid., p. 229.

30. As Marx says in his *Grundrisse*.

31. This is similar to the problem inherent in Kuhn's view of scientific theories and practices as paradigms. See his *The Structure of Scientific Revolutions*, Chicago 1962.

32. K. Marx & F. Engels, *The German Ideology*, in R. Tucker (ed.), *The Marx–Engels Reader*, N.Y. 1972, p. 13.

33. *For Marx*, p. 231ff.

34. See K. Marx, *Economic and Philosophic Manuscripts of 1844*, in K. Marx, *Early Writings*, Harmondsworth 1977.

3

On Oppositions to Reductionism
Hilary Rose and Steven Rose

It is not enough to know that we are "against" reductionism, like sin; we have to be clear *what* it is we are against and *why* if we are to move forward. The framework for the present debate can best be understood by locating it historically. This short paper was written partly to clarify our own minds about some of the issues involved and also to act as signpost to some of the current literature which seems relevant to the discussion which follows later in the book.

The meaning of science

Science is a term used to describe necessary ways of knowing the material world, of explaining and theorizing it as a guide to action. There are many different "ways of knowing" which have been developed in different societies and at different times, and within our own society by different social groups. The questions that confront us are about how to find emancipatory, liberatory ways of knowing. Because the goal is a society based on that old materialist ideal, the freedom of necessity, liberatory ways of knowing are those which are most in accord with the materiality of the world.

It is here that the contradictions of modern science, like those of capitalism, lie. Initially a destroyer of medieval cosmology, it acted as a midwife to capitalism. At the same moment it contributed both toward the hegemony of the bourgeoisie through the material forms of its new technologies and weaponry and served as a force of production in the central process of capital accumulation. The new science was socially progressive in its liberation of human thought from the shackles of the old cosmology. In a manner distinct from all earlier forms of knowing the world, it had to generate predictions and technologies that "worked" because the capitalist mode of production now depended upon the success of these predictions. The dominant mode of explanation (which was finally to triumph in the nineteenth century) was that of mechanical materialism, reductionism. Scientists saw themselves and their reductionist method as socially progressive, on the side of human liberation (engineers, more intimately linked to the development of the forces of production, had a different, less critical history, but that's another story).

Within capitalism, therefore, science offered a way of knowing the natural and social worlds which would guide socially progressive practice. Hence the revered M&E claimed to be doing a "scientific socialism" (we don't wish to enter into that debate at this moment, however). Today such claims and theories are in crisis. The crisis is of several orders. The authority of science is seen as an aspect of bourgeois hegemony, a domination of nature and humanity based on a rationality which threatens to destroy us beneath the wheels of a scientific and technological juggernaut. The "objectivity" of science is seen to dissect fact from value and delegitimate our own subjectivity. At the same time, within science itself, despite the still enormous power of the reductionist paradigm, there are signs of insoluble problems and irreconcilable paradoxes (high energy physics, ecology, mind/brain relations, developmental biology, evolutionary theory, are examples, the last three particularly being the topics of our discussions here).

Responses to the crises

There are a variety of responses to these crises, based on differing interpretations of science and of scientific knowledge and its relation to the material world.

(1) Within capitalism, a naive scientific positivism/realism still argues a single science. Reductionism, if it is to be faulted, is merely bad methodology. Thus scientific theories are regarded as successive approximations to "truth"; today's science is more "truthful" knowledge about the world than yesterday's, etc. The failure of sociobiology, etc. is thus seen, in general, to be a failure by its own rules. It is "bad science" which makes unwarranted assumptions and uses inadequate evidence and poor statistics – e.g. in IQ theory, cognitive sex differences, etc. The critique of reductionism shows how it violates its own norms, thus destroying it without needing any fundamental recasting of scientific theory. This can be done from a "purifying bourgeois science" perspective – as in the work of Leach or Medawar[1] – or as specifically "critical science" – for instance Fairweather, Kamin, Montagu, Sahlins.[2]

(2) Since the 1930s in post-revolutionary societies and "orthodox" marxism, there have been a number of overlapping, sometimes simul-

taneously held, and often contradictory positions in science. Thus, it
has been maintained: (a) that doing science is by definition progressive; the methodology of bourgeois science is in contradiction with the
constraints which bourgeois society places upon both production and
knowledge, and science therefore overthrows capitalism (Bernal, Mil-
lionshchikov, Mandel[3]); (b) that scientific knowledge has clear social
determinants which affect its form, history, and possibly knowledge
structure (Hessen, Bernal, and the subsequent externalist history of
science tradition east and west[4]); (c) that it is possible to build a specifically socialist science which will know the world differently and
without contradiction, e.g. the attempts to apply dialectical ontology
to science in the Soviet Union in the 1930s and by Lysenko[5] – *theory
itself* was shaped as knowledge and class perspectives/practice, while
material practice and the democratization of science was the thrust of
the Cultural Revolution in China;[6] and (d) this last position was given
a theoretical framework as the "orthodox" dialectical materialism of
the 1930s (see Graham on the Soviet Union, or Haldane or Prenant in
Western Europe).[7] In these accounts of socialist science, it is argued
that nature does not work in a reductionist but in a dialectical manner.
Reductionism is a product of a bourgeois world view. Marxist science
will build a socialist society and act as the physician for bourgeois epistemology. The dialectic may be found in nature, and we understand
nature by means of the dialectic.

(3) Over the last decade, within capitalism, socialists have occupied
all of these positions, and laid siege to bourgeois hegemony from
within them. Thus reductionist methodology *"done properly"* has
been used by Kamin, by those arguing against recombinant DNA or
dangerous technologies such as nuclear power, and in union struggles
over health and safety issues. The "people's experts" can beat the
enemy at their own game.[8] *Theory itself* has been attacked by the
women's movement, Science for the People, etc. *Changed material
practice* has been part of the struggle of the women's movement, radical collectives, etc. The *dialectics of nature* has been revivified by
Gould, Levins, Dialectics Workshop and the School of Marxist Education in the U.S.[9] There is a rebirth of a sort of Hegelian structuralist
dialectics by radical Piagetians, and some other developmental and
cognitive schools in Europe.[10]

(4) One of the consequences of these challenges has been a steady
reconstruction of the bourgeois philosophy of science, in which
"truth" has either disappeared completely or has gone relative, historically or totally.

Bourgeois philosophy and relativism

(1) Bourgeois philosophy of science is in a strange crisis; after Lakatos and Kuhn and their opening up of the Popperian world,[11] the claim that science is about "truth" has been largely abandoned for talk of "research traditions" and "rationality" in theory-making. Theories are ways of integrating facts (sometimes but not always seen as "value-free") in a coherent way which provides rich puzzles to solve. Choices between research traditions may be rational; e.g. increasing anomalies, "degeneration", lack of "richness" (Lakatos); or may depend on individual psychology during periods of revolution (Kuhn); they may be related to complex factors including the prevailing "world view" and its compatibility with particular theories (Laudan).[12] In any event the real world is abandoned in favour of rational crossword puzzles. Attacking reductionism from within this framework would be to insist on its replacement as a theory structure for some of the reasons cited above; the battle would be in the cognitive domain, and would have to show that dialectical science could overcome reductionism's anomalies, etc. So far as we know, however, most of this group of philosophers of science have been favourable to reductionism/sociobiology. Perhaps this is not accidental. They include, for example, Flew, Urbach, Ruse[13] and, to some extent, Laudan.

(2) *Bourgeois relativism.* Exemplars include Feyerabend[14] (nearly but not quite – see H. Rose[15]). Most conspicuous are the Edinburgh school, Barnes, Shapin and Bloor.[16] In this analysis, criteria for choosing research traditions are removed from the cognitive domain and instead are determined by "interests"; all knowledge is relative and itself the object of study by anthropologists and sociologists of knowledge – but on what ground do they stand when they make their own study and claim its scientificity?[17] Laudan also makes this point. Bourgeois relativists have, of course, no position on reductionism versus dialectics except as an exemplar in which they could perhaps search for the conflicting interests involved; e.g. sociobiologists tend to be WASP, dialectical materialists, in the US at least, Jewish(?). Such a social perspective is essentially static, despite the phenomenological stance. The crises that this stance generates are discussed below.

(3) *Left relativism.* This is the "science is social relations" position of Young and others of the Radical Science Journal collective.[18] In this analysis science is an aspect of a bourgeois world view determined

by class perspective. Bourgeois science is the ideology of commodity-fetishism of which reductionism is but one expression. One opposes bourgeois science by a working-class science, people's science or feminist science (though they don't say the latter – it has been left to Stéhèlin and Haraway to raise that question[19]). This socialist science would be a different way of knowing the world which is liberatory *because* it is in accord with the interests of oppressed strata. That is, all we ultimately know is the social world; the social and economic base determine our science, and as we change the base so we make a new science. It is therefore enough to expose the bourgeois nature of reductionism to dismiss it, but any critique from within the knowledge structure itself is *part* of an ideological mystification (e.g. Levidow, Figlio[20]). When we make our new society, which will be socialist, non-patriarchal, etc., we will have a new and qualitatively different science simply by virtue of the social change. Replacing reductionism with a dialectical science is in this view an expression of the class base and partisan nature of our science (for a critique of this position see[21]).

Toward an emancipatory science

The tasks of an emancipatory science are defined by the critique of contemporary bourgeois reductionist science. It must overcome and transcend: (1) the subjectivity/objectivity split; and (2) the domination of the natural and human worlds by instrumental rationality. And it must achieve: (1) the democratization of science; and (2) a dialectical view of nature, including human nature, as neither static nor infinitely plastic; that is, a materialist view.

In accord with this, of course, our own present position is not static but has undergone transitions beyond our earlier writings– e.g. when Martin Barker quotes from *The Conscious Brain* it feels like a bit of prehistory. So the following must be regarded as a way-station, not a definitive statement.

Ways of knowing the real world are not infinitely flexible, all equally plausible or merely relative. They can be better or worse and can and must be judged by how far in accord with the "material world" they are (yes, of course, we *alter* that world in some measure by our views of it– it is a world becoming, not a world static– see below– but it is still not infinitely alterable and how it is altered is itself rule-dependent). We do not have isolated facts to be ranked and sorted by value/socially laden theories; fact and theory are intimately intertwined; yet it is possible to know some positive and some negative things about the world. In the Popperian sense, theories *can* be disconfirmed; e.g. the theory that we can change the weather by praying – Dick Lewontin's example. The belief that the earth is approximately spherical, that there are moons of Jupiter, and that DNA is a double helical structure which carries genetic information, are claims more in accord with the arrangement of the material world than that the earth is flat, that Jupiter has no moons or that protein carries the genetic information. Our knowledge is thus historically relative in that we judge it by (among other things) knowing that it is "more the case" than what went before.

Against scientific positivism, we argue that the historical world view of the observer cannot be separated from the facts/theories the observer subscribes to. Scientific knowledge is historically relative, and criteria for believing/not believing things are not entirely within the internal cognitive domain of a particular field. Criteria of concordance with other parts of a world view, including social (and aesthetic and religious) parts, are clearly relevant. The critical analysis of the claims of reductionism according to its own norms is *part* of the fight against oppressive science but is strictly limited. For instance, the response to Kamin's or Fairweather's critique of IQ or cognitive sex difference theories by mainstream reductionism is to do "better controlled" studies – as Jensen *et al.* now claim, or to attempt to "root" the claims more deeply in biology – e.g. Eysenck and the Hendricksons. Critical stances within the reductionist problematic can only be one, tactical part of a battleground (though left relativist attempts to dismiss those fighting on this terrain as in fact part of the enemy are divisive, and to put it at its most charitable a mistake; some would put it more strongly – see[22]). Thus our attack cannot merely be on "bad science", though the concept of "bad science" repays analysis.[23] If one is committed to claiming that a large part of what is called "science" in a particular terrain is "bad", or "ideology masquerading as science" (e.g. as we ourselves have done, following Althusser, in the past), there is likely to be more going on!

In its critique of the naiveté of scientific positivism, bourgeois mainstream philosophy of science is surely justified to point to the shakiness of its understanding of terms such as the "truth" about the material world, etc. But in order to get to this position, it has had essentially to insist on a technology/science, fact/theory distinction; it does not allow use of the criterion of practice to test a theory. Any theory, it claims, can fit any practice, and practices are themselves value-free; it is only theories which are about rationality, and not truth. Going back to the previous point: the advances of science are not merely telling us different things about the world but things which are more in accord with its materiality. Einsteinian relativity theory or the DNA theory of heredity or the neuronal theory of brain function are *more* like the material world than Newtonian mechanics, Darwinian gemmules or syncytial theory. The Laudanian or Lakatosian assertions are part of a bourgeois mystification of the world that may suit bourgeois rationality. Feyerabend is right in one respect; if forced to choose between realism and rationality, science must choose realism, and so must socialists and revolutionaries, as Bhaskar[24] and Benton[25] argue from the left.

The arguments against bourgeois and left philosophical relativism develop *a fortiori* from the preceding paragraph (and we've discussed them elsewhere[21,22]). Bourgeois relativism explicitly claims to disarm the critics of bourgeois science (as merely another interest group); it is both *against* any science except its own conception of sociology, and *for* the status quo. Left relativism is more serious because it offers the promise of a revolutionary critique while actually disarming the left; our choices of what to believe about the world are neither infinitely plastic nor so mechanically linked to our class position that no change in consciousness and action is possible (unless free-floating intellectuals are in a Mannheimian way above class!). Practice on this left relativist model becomes essentially voluntaristic, an expression of personal commitment to a "serious" belief about the world. Meanwhile, inside Jericho, reductionism triumphs.

The problems for classical dialectics are multiple. Its rigidity is not merely accidentally associated with stalinism and the vulgar marxism of the "iron laws of history". There is a simultaneous appeal to scientificity as being beyond and above mere human practice and an insistence on the existence of rigid ontological categories into which the natural world must be fitted; but human knowledge of these categories is itself seen as unproblematic. The Lukàcs/Korsch critique of Engels's ontologizing dialectic was to relocate the dialectic as lying between humanity and nature and in the social world, and is germane here. We also need to consider how far the ontologizing of the dialectic is part of the "domination of nature", of which many would argue that marxism has to date been as guilty as bourgeois science.[26]

So we would argue something like the following. The development of scientific knowledge of the world is cumulative and in one sense progressive. The victory of capitalism over feudalism *did* liberate the human spirit and result in a way of knowing which was better in the sense of being more in accord with the material world. The programmatic of capitalist science is reductionism, which is a powerful but flawed tool for studying and knowing the world. It carries with it contradictions which, as science and capitalism develop, turn it from being liberatory to being oppressive, from knowledge-bringing to knowledge-restricting. In a period of intense struggle over competing world views, the inadequacies of reductionism, like those of capitalism, generate opposition which seeks either to oppose scientific rationality entirely[27] or to replace reductionist rationality with alternative rationalities at least some of which contain an emancipatory project – e.g. systems theory, structuralism, dialectics. But any of these tentative alternatives to reductionism are themselves historically relative. We have neither absolute ontological nor epistemological yardsticks.

A new science as it develops will not destroy the genuine insights of reductionism, nor its power as a methodological tool in research (Levy's "method of isolates"[28]), nor will it endeavour to cut humanity adrift from our biological natures. The strengths of reductionism will be incorporated into a post-reductionist science, its limitations and errors transcended. In certain areas reductionist explanations will be seen as limited special cases (for instance within-level, temporally organized, closed systems). In many, such explanations will be bypassed as relating to trivial non-problems arising from bourgeois ideology (for instance, mind/brain reduction, phenotype/genotype reduction).

Working out the relationships between the different sciences, knowing as variables what bourgeois science sees as constants, emphasizing the historicity of objects and the reality of discontinuities,[29] is to work towards both a transformative practice and a transformative theory of the natural world.

Notes and references

1. For instance, E. Leach's reviews of the hereditarian literature in *The Listener*; P. B. Medawar and J. S. Medawar, *The Life Science; Current ideas in Biology,* London 1977.

2. H. Fairweather, "Sex differences in cognition", in *Cognition,* no. 4, 1976, pp. 231-80. L. J. Kamin, *The Science and Politics of IQ,* Harmondsworth 1979. A. Montagu, *The Nature of Human*

Aggression, Oxford 1979. M. Sahlins, *The Use and Abuse of Biology*, Michigan 1979.

3. J. D. Bernal, *The Social Functions of Science*, London 1939. Anon., *The Scientific and Technological Revolution*, Progress Publishers 1972. E. Mandel, *Late Capitalism*, London 1976.

4. For B. Hessen's article, see *Science at the Crossroads*, Kniga 1931; J. D. Bernal, *Science in History*, London 1965.

5. For an account of Lysenko, see D. Lecourt, *Proletarian Science?*, London 1978.

6. For an account, see Science for the People Collective, *China: Science Walks on Two Legs*, New York 1974.

7. L. Graham, *Science and Philosophy in the Soviet Union*, New York 1972. J. B. S. Haldane, *Dialectical Materialism and Modern Science*, Labour Monthly Pamphlet, 1942, and his introduction to *The Dialectics of Nature*, London 1940. M. Prenant, *Biologie et Marxisme*, Paris 1936.

8. For examples, any issue of the BSSRS *Hazards Bulletin*, or Science for the People's work on recombinant DNA in Boston and Cambridge, Mass.

9. For instance the journal edited by H. Tarkington, *Science and Nature*, and the discussion documents put out by the School of Marxist Education in New York.

10. See the journal edited by J. Mehler, *Cognition*.

11. This debate is enjoined in I. Lakatos and A. Musgrave, *Criticism and the Growth of Knowledge*, Cambridge 1970.

12. L. Laudan, *Progress and its Problems*, California 1977.

13. A. G. N. Flew, *Evolutionary Ethics*, London 1967. R. Urbach, "Progress and degeneration in the IQ debate", in *Brit. J. Phil. Sci.* no. 25, pp. 99-135 and 233-59, 1974. M. Ruse, *Sociobiology: Sense or Nonsense?*, Dordrecht 1979.

14. P. Feyerabend, *Against Method* and *Science in a Free Society*, London 1975, 1978.

15. H. Rose, "Hypereflexivity: a new danger for the countermovements", in H. Nowotny and H. Rose (eds.), *Countermovements in the Sciences*, Dordrecht 1979, pp. 277-90.

16. B. Barnes and S. Shapin, *Natural Order*, London 1979; B. Latour and S. Woolgar, *Laboratory Life*, London 1979.

17. H. Rose and S. Rose, "Against an Oversocialized Conception of Science", in *Communication and Cognition*, no. 13, 1980, pp. 173-87.

18. See the last few issues of *Radical Science Journal* and especially R. M. Young, "Science in Social Relations", in *Rad. Sci. J.*, no. 5, 1977, pp. 61-131.

19. L. Stéhèlin, "Sciences, Women and Ideology", in H. Rose and S. Rose (eds.), *The Radicalisation of Science*, London 1976, pp. 76-89. D. Haraway, "The biological enterprise: sex, mind and profit from human engineering to sociobiology", in *Rad. Hist. Rev.*, no. 20, 1979, pp. 206-37.

20. L. Levidow, "A marxist critique of the IQ debate", in *Rad. Sci. J.* no. 6/7, 1979, pp. 13-72. K. Figlio, "Sinister Medicine? A critique of left approaches to medicine", in *Rad. Sci. J.* no. 9, 1980, pp. 14-68.

21. H. Rose and S. Rose, "Radical science and its enemies", in R. Miliband and J. Saville (eds.), *Socialist Register*, London 1979, pp. 317-34.

22. See the entire issue of *Science Bulletin*, no. 22, 1979.

23. R. C. Lewontin, S. Rose and L. Kamin, *Biology, Ideology and Human Nature* (forthcoming).

24. R. Bhaskar, *A Realist Theory of Science*, Hassocks 1978.

25. See T. Benton and others (e.g. K. Soper) in J. Mepham (ed.), *Issues in Marxist Philosophy*, vols. 1-3, Hassocks 1979.

26. A. Schmidt, *The Concept of Nature in Marx*, London 1971. W. Leiss, *The Domination of Nature*, New York 1972. M. Bookchin, *Post-Scarcity Anarchism*, London 1974.

27. H. Nowotny and H. Rose (eds.), *Countermovements in the Sciences*, Dordrecht 1979.

28. H. Levy, *The Universe of Science*, New York 1932.

29. R. C. Lewontin and R. Levins, "The Problem of Lysenkoism", in H. Rose and S. Rose (eds.), *The Radicalisation of Science*, London 1976, pp. 32-64, and R. Levins, *Marxism and Natural Science*, unpublished MS.

4

Cleaving the Mind: Speculations on Conceptual Dichotomies

Lynda Birke

I felt a Cleaving in my Mind –
As if my Brain had split –
I tried to match it – Seam by Seam –
But could not make them fit.

The thought behind, I strove to join
Unto the thought before –
But Sequence ravelled out of Sound
Like Balls – upon the Floor.
Emily Dickinson, 1896.

The main theme of this conference is to move away from reduction-ism, to pursue a more dialectical understanding of the concept of mind; and to this end, we are reminded of the futility of continuing debate about the horrors of reductionist thinking. On the other hand, as Martin Barker points out in his paper for this conference, we need to understand *what* we dislike about reductionism if we are to move beyond it. Simply bringing fifty or so people together does not guarantee that they will agree what it is they don't like.

With that in mind (if you will forgive the pun), I want to bring into the discussion of reductionism a related question: that of seeing the world in terms of simple categories, and in particular, in dualities (and by duality, I mean here more than simply mind-body duality). I real-ize that we have to parcel the world up in some way, to categorize it, if we are to begin to understand it. As an ethologist, for example, I have to contend with the problem of how to define "chunks" of an animal's behaviour in order that I can understand why the animal does what it does. Are they only chunks which I perceive (or which I can easily program into a computer)? Or are they, in addition, chunks which

60

are meaningful in the life of the animal? And do these chunks which I try to define, only valid in a particular situation, become quite meaningless in others? But somewhere, somehow, we have to draw boundaries. The difficulty, I think, is that we tend to draw static boundaries, and then to start the reductionist process within those boundaries, forgetting how arbitrary they were in the first place.

All human societies use some scheme of classification to order perception of the world: indeed, some classificatory schemes seem to be used, in some form or another, by all known societies (for example, the use of binary classification discussed by Needham,[1] and the discussion later in this paper). A second difficulty for our understanding of the process of reductionism, then, is the extent to which we understand how the "commonsense" classifications which we use in our own society inform, or even determine, the classifications which we establish in the process of generating hypotheses about the natural world.

As I have indicated, I am primarily concerned with dualities here, with boundaries between pairs of opposites. I am particularly concerned with the dualism of gender, although I have adopted a rather eclectic approach to it. I am not offering this paper as any kind of "definitive" statement, but very much as an exploration, aimed at kicking off some discussion. I hope it will be taken in that spirit. I also wish to stress that, by focusing on particular examples of dualities, I am not suggesting that these represent the only examples of the ways in which classifications can contribute to the reductionist process. It is not, therefore, a paper "about" gender polarities, but uses these to address questions of the ideological role of reductionist science.

Dichotomous, and rigid, categories plague my life in several ways. For a start, they plague me as a biologist interested in hormones and behaviour, and they plague me as a feminist when I have to think/write/teach, and have to use concepts such as gender. This is a field which is full of separation into *either* one category *or* another. I would suggest that this process of separating off into mutually exclusive categories can be an important first step in reductionism. Suppose we consider, say, "explanations" of women's under-achievement in the sciences in terms of brain lateralization, or biologically based "explanations" of racial inferiority on IQ tests. I assume that we all agree that these are examples of reductionist ideology at its most transparent. What I want to emphasize, though, is that the first point of this conceptualization is that people are being divided up into two categories, one group of which are likely to have their behaviour explained by reduction to lower, biological, levels. This is partly a product of the empiricist tradition of Western science, which seeks to discover universal, natural laws:[2] from this, those individuals who do not conform to the laws are defined as deviant, requiring special explanation. And partly it is a product of a particular ideology: it is noteworthy that it is usually the groups accorded less power and status in our society which have their behaviour thus explained.

The gender dichotomy as a case-study

> There is perhaps no field aspiring to be scientific where flagrant personal bias, logic martyred in the cause of supporting a prejudice, unfounded assertions, and even sentimental rot and drivel, have run riot to such an extent as here. [1910][3]
>
> [The explanation of female irrationality] is to be found in all the physiological attachments of woman's mind: in the fact that mental images are in her over-intimately linked up with emotional responses . . . that intellectual analysis . . . involves an inhibition of reflex responses which is felt as neural distress . . . and that woman looks upon her mind not as an implement for the pursuit of truth but as an instrument for providing her with creature comforts in the form of agreeable mental images. [1913][4]

We might smile at these quotations today. The trouble is that the first statement is still largely true. I am not going to bother you with the countless recent examples of the latter type of statement, nor am I going to debate in detail the question of the validity of the sex-differences research itself, as that has been done elsewhere.[5] I am going to do no more than assert that on the basis of these studies, I believe that gender has very little to do with biology, and is a social construct. The literature on sex differences is vast, and I can do no more than refer potential dissenters to the review papers listed under reference 5.

What I want to do, however, is to approach gender dichotomies from a different angle, namely the literature on human sexuality. I do this partly because it is a field with which I am (all too) familiar, and partly because it is less commonly thought of as a "problem", even by critics of biological reductionism, than, say, sociobiology. I think it is important to look at some of the problems involved in a specific area, in order to see that simple dichotomies don't get us very far.

Among other things, my research has led me into the sordid realms of theories of the aetiology of homosexuality. Now it may seem legitimate to some to ask such specific questions as: can we isolate factors which are clearly correlated with (causal of) homosexual development? From there, one might look at "biological" variables, such as prenatal hormones, or "social" variables, such as Over-Intimate Mothers.[6] The legitimacy of such research, however, rests on the assumption that we can quite simply bisect the human population. Then we can conceptualize one set of people as normal, and the other as some sort of problem. "Problem" people are commonly thought of as being closer to nature: indeed, "to think otherwise would be to call in question modes of representation conceived as universal".[7] As a result, problem people, the objects of scientific study, are not normally white, middle-class, heterosexual, men.

The significance of the initial dichotomy, then, is that it helps to establish which people are deemed worthy of study. It is noteworthy that it is only in the last couple of hundred years that, for example, women have been seen as a "problem",[8] and that "the" homosexual has emerged as a social and medical anomaly. Before that, homosexuality was seen as a sinful potential in all of nature: something to be avoided and judicially punished, to be sure, but a potentiality in all of us.[9] It is, therefore, quite recently that homosexual people have been defined as a problem,[10] which has increasingly been seen as a *scientific* problem, steeped in reductionism. "Homosexuality" might be reduced to genetics, or to hormonal "aberrations". The latter are currently exciting much attention in some quarters as they promise opportunities to "prevent homosexuality" by injections to the pregnant mother.[11]

As the initial dichotomy determines who is studied, then it is less than surprising that dichotomous categories are common in this literature. Important dichotomous categories include biological sex, gender identity, social sex role, and sexual orientation. These are different, and potentially separable, categories,[12] although one could be forgiven for thinking otherwise after reviewing the field. Each is commonly conceptualized and used as generating polar opposites: biological sex is female *or* male; gender identity involves believing oneself to be female *or* male; social sex role involves femininity *or* masculinity; while sexual orientation is homosexual *or* heterosexual. Each concept presupposes that people can be neatly divided into one or the other state, and leads to the supposition that simple causal factors might be found.

This is the trouble. None of the categories mentioned above *is* in practice composed of easily defined opposites. Obviously, biological sex comes closest: there are men and there are women. But sometimes these apparently simple distinctions break down. How do we define sex? The distinction to most people is based on the appearance of the external genitalia, while to a biologist, the primary distinction is likely to be chromosomal – XY is male, and XX is female in mammals like ourselves. But having only two categories presents us with problems of definition when we find individuals whose sex chromosomes do not fit this simple classification (as in cases in which the chromosomes have not separated in the usual way during cell division, resulting in, for example, XXY, X, XXX, XYY etc.). We face even bigger problems when we find individuals whose chromosomes say one thing, while they appear to be another. This is the case, for example, with congenital androgen insensitivity. These are males, in that they are XY, and they possess testes which secrete androgens. They do not have, however, the enzymes which normally make use of androgens in the tissues, so their tissues remain insensitive to it during foetal and later life. They therefore do not become externally masculinized, and appear to be female, and so are brought up accordingly. And how do our concepts of biological sex fare when we discover humans who appear to be born female, and miraculously change sex at puberty as a result of a rare metabolic disorder?[13] Of course, people with indeterminate phenotypic sex are in big trouble: they are born into a society which classifies people as one sex or the other, and which attaches a great deal of symbolic value to these categories. It is not surprising that so much importance is now attached to attempts to fit people into the two categories by "corrective" surgery, hormone injections and so on.

When it comes to the next category, gender identity, there are similar problems. Clearly, the vast majority of people think of themselves as unequivocally one or the other sex. In most cases, this is simply in accord with an unambiguous biological sex. As before, though, there are people who don't fit. Trans-sexualism is one example, since these are people who feel that the gender of their "mind" is not congruent with that of their "bodies". Gender identity is not a product simply of biological determination, as many studies have shown: much depends on the sex to which an individual is assigned at birth.[14]

It is in the last two categories that the biggest difficulties arise. Social sex role refers to the characteristics of individuals that are culturally associated with one or the other sex, characteristics such as physical appearance, mannerisms, personality, and so on. It is often supposed that these represent fairly immutable characteristics, with femininity and masculinity being seen as opposite poles, and often describing different people. The fundamental difficulty is that this dichotomous classification is itself rooted in a biological classification (male versus female), so that deviations from the opposed stereotypes are often assumed to be based upon this biological division, an assumption which then determines the hypotheses set up in research. Furthermore, these distinctions permeate the homo-/heterosexual dichotomy. Among gay people, for example, it is those who more obviously flout the socially accepted criteria of behaviour appropriate to one's biological sex who are likely to be labelled as homosexual – the effeminate male/butch dyke stereotypes. Those whose behaviour is more in accord with stereotypes appropriate to their biological sex are not so readily labelled.[15] Behaving in these ways is more likely to be a response to social expectations of homosexuals than it is to be a function of some biological variable.[16] Nevertheless, the literature tends to assume that gay men are less "male" than other men, while gay women are less female or more male. It is this assumption which lies behind the kind of questions asked in research, questions such as whether gay men have less of the "male" hormones (androgens) at some point in their lives. Once hormones have been defined as "male" or "female" (which is inaccurate in any case), and once gay people have been defined as less male, less female, etc., then the hunt is on for indicators of lesser maleness or femaleness, such as smaller quantities of androgens. Fried[17] analyses the use of language in shaping how we view the world of gender, and comments on the assumption made by Money and his colleagues that dichotomous classes naturally exist. She quotes a section from Money and Ehrhardt[18] which illustrates the point:

> Actually, such a person (gay man) has an identity/role that is partially masculine, partially feminine. The issue is one of proportion: more masculine than feminine. Masculinity of identity manifests itself in his vocational and domestic role. Femininity of identity appears in his role as an erotic partner; it may be great or slight in degree and it is present regardless of whether, *like a woman*, he receives a man's penis, or, also *like a woman*, he has a man giving him an orgasm. It goes without saying that the ratio of masculinity to femininity varies among individuals (emphases mine).

Note how the equation is made between sexual acts defined as receptive/passive and femaleness/male homosexuality. Fried comments on this:

No – it only goes with saying, and saying, and saying that there exist such polar concepts as masculine and feminine, which, placed in varying ratios to each other, form the bedrock of our personality. It only goes with saying that a man who chooses to have sex with another man has a feminine identity as an erotic partner. We could just as easily name it spiritual, or bestial, or classically Greek, or anything else we like, and thereby alter our interpretations of the experience to conform to our pre-conception of it. But the masculine-feminine duality in behaviour is the reigning construct of Western civilization. So, finding it difficult to fit male homosexuality into their conception of the "masculine" role, Money and Ehrhardt must therefore assign it to the "feminine" role – a strange solution, one would think.

Within its own terms, some of the research might be said to be moving slightly from rigidly bipolar distinctions. Concepts of masculinity and femininity have for years been founded upon assumptions of their opposite nature: psychological tests have been founded upon this assumption.[19] Recently, however, there have been pleas to consider masculinity and femininity as separate dimensions[20] rather than as opposites, so that a person might be thought of as having both "masculine" and "feminine" attributes. Even the biological changes thought to underlie sexual differentiation are becoming less clearly one-dimensional. Once thought of as a single process of masculinization, it now seems that at least two processes (masculinization and defeminization) are involved in making males.[21]

But these subtle changes do not challenge the basic dualism and its assumptions. They mean we might have to think a little harder about masculinity and femininity, but we retain the essential dichotomy, and we continue to root it in our dualistic conception of biological sex:

The fundamental problem with accepting *a priori* the sexual duality as the primary construct of reality is that all of our discussions about sex and gender must then take place *within* this construct. All that the recent work on the relationship of sex to gender has done is to remove us from consideration of human activity in two categories (male and female) to one in two pairs of categories (male, masculine; female, feminine). And we are left to spend our time squabbling over whether each trait displayed by a man is more rightly attributed to his maleness (sex) or to his masculinity (gender).

Dualistic thinking, then, can present us with many problems. I will return to these problems, and the question of the functions served by perpetuating these dichotomies, later. So far, I have only referred to gender as a way of making divisions. There are, however, many others, many of which are used as global constructs in the way that gender is, and which are frequently linked conceptually with gender. The next section, then, is a sketch to show some of the associations made.

The world in two halves

"Femininity" and "masculinity" are clearly not the only pairs of opposites by which we classify our world. Many other paired concepts are used, some of which are explicitly associated with gender. That is, one of the pair is commonly conceptualized as inherently feminine, or incorporating features we define as feminine, while the other is masculine. This conceptualization is by no means unique to Western philosophical traditions: seeing the world in binary opposition would seem to be a near-universal feature of human societies.[22] What is of interest is not so much the opposition itself, but the prevalence of certain associated paired concepts and the values we attribute to them. For example, Star[23] notes the distinctions between popular concepts of left and right, as shown in a test given to college students, and she cites the following:

> The Left was characterized as bad, dark, profane, female, unclean, night, west, curved, limp, homosexual, weak, mysterious, low, ugly, black, incorrect and death; while the Right meant just the opposite – good, light, sacred, male, clean, day, east, straight, erect, heterosexual, strong, commonplace, high, beautiful, white, correct.

Similarly, Ornstein[24] divides concepts along the following lines: Masculine equals Yang (of Chinese taoism), day, intellectual, active, analytic, right (of body), left (of brain), sequential, focal, light, time, verbal, causal, logical argument; while Feminine is Yin, night, sensuous, receptive, Gestalt, left body/right hemisphere, simultaneous, diffuse, dark, space, acausal, experiential. Femaleness, or femininity, is conceived of by many human societies as dark, or weak, terms which tend to have negative connotations, at least to us.

While some writers simply imply associations between paired concepts, others emphasize the social and ideological significance attached to these associations.[25] For instance, aggressive masculinity has been attributed especially to particular groups of people as a means of justifying their hegemony. Hoch, and others,[26] have implied that the cult of masculinity and "masculine" values has characterized the recent history of Western civilization, and in particular, capitalist society. Hoch further suggests that "masculine" values have often been employed to justify imperialist expansion and racism, people of other races being viewed as "soft" and feminine, less able to look after themselves:

> The Anglo-Saxon upper classes of England and America were proclaimed to be the unique possessors of a manly valour which entitled them to rule over lower classes and races characterized by degeneracy and effeminacy. "The Germanic race, and especially the Anglo-Saxon part of it," wrote the American historian Francis Parkman, "is particularly masculine, and therefore peculiarly fitted for self-government."[27]

Sometimes the association between other races and "femininity" was more explicit. Hoch refers to the long history of anti-semitism, and to the medieval myths which maintained that Jewish males menstruated like women.[28] We can discern a similar opposition between the masculinity of the ruling Caucasian and the less desirable "femininity" of others in a statement made in 1906.

> The Caucasian . . . is dominant and domineering, and possessed primarily with determination, will power, self-control, self-government . . . and great reasoning powers . . . The negro is primarily affectionate, immensely emotional then sensual, and under provocation, passionate . . . They are deficient in judgement, in the formulation of new ideas from existing facts, in devising hypotheses, and in making deductions in general.[29]

The epitome of masculinist ideology is, according to some writers, fascism.[30] The espousal of "masculine" values is very evident in the writings of the Nazis, and, more recently, in the publications of the National Front. Much is made in these publications of the need to fight the "effeminacy" of liberalism, or worse (!) of marxism.

It seems, then, that a tendency to see the world in two halves, and to consign these halves to gender categories, is quite a common affair, and reminds us of the significance which we attach to what seems like a rather innocuous distinction between the sexes.

There are, of course, other dualistic concepts which are not overtly related to gender dichotomies: I have simply focused on those which are so related. The concept of discontinuity in evolution might be thought of in this way. Cartesian dualism itself is not entirely free from associations with gender, in that men were thought to be more associated with the mind, which became, therefore, a masculine preserve, while women were held to be closer to nature and to their bodies. Thus, in Victorian times, women were told that the development of their minds through study would cause untold havoc, making them weak and vulnerable.[31]

Although there are clearly myriad ways of viewing the world as dichotomies, I have concentrated on those which carry gender connotations, precisely because of the ideology attached to them. The problems arise not so much because of the association with gender as categories, but because the gender dichotomy so often becomes elevated to the status of an explanatory variable, as we saw in the case of the literature on homosexuality. What appears at first to be a purely linguistic association with a gender term becomes instead an organizing principle around which research is conducted.

Some questions (and answers)

Various suggestions have been made as to why we see the world in gender-associated dualities. It might, for instance, be construed in Hegelian terms as a distinction between Subject/Other (as articulated by Sartre in *Being and Nothingness* and, more explicitly, by De Beauvoir in *The Second Sex*). For De Beauvoir, the sense of otherness begins with the child's separation from its mother: she maintains that women retain the status of "other" as they subsequently learn their role in society, while men learn to accept their status as subject. But there are problems with this view. Apart from the deep-rooted misogyny threading through some of the writing in this genre[32], existentialism is basically pessimistic and individualistic. If the gender duality results, ultimately, from the first separation of a child from its mother, then what hope have we of moving beyond it? There will always be *someone* who becomes cast in the role of perpetual Other. Existentialism can be criticized as an individualistic philosophy *par excellence,* as Aronson points out:

> Individualism is a logic of despair. So is Sartre's unhistorical and ontological thinking. Problems inscribed in *being,* be it individual or social, are problems for all time. If history enters the analysis thereafter it is already too late: all the traps have already been set.[33]

Seeing gender duality in terms of subject/object distinctions cannot help us much, as it ignores any historical change in the significance accorded to the duality. And the universal nature of this distinction implies little hope of changing anything.

A second possibility put forward by some feminist writers is that our tendency to see the world as gender polarities is associated in the

West with the rise of monotheistic patriarchal religion. This, it is argued, espoused the values which we now designate as "masculine", and suppressed with force any other cults or religions. As it is described, it would seem that the rise of an autocratic father god was indeed a bloodthirsty affair, and the suppression of earlier "pagan" religions took far longer than we are led to suppose, lasting well into the eighteenth century.[34] We may or may not accept these suggestions as based on historical "fact"; but they do bring out one interesting point. While they base the gender duality on a concept, patriarchy, which is notoriously difficult to define,[35] some of these writers do suggest that concepts of gender have changed in significance in the recent past. Singer,[36] for example, notes that our present notion of gender embodies both a rigid separation of the two opposites, as well as implications of superiority/inferiority. She suggests that other philosophies which embody an idea of gender do not have the same rigidity or hierarchy. She discusses, for example, some earlier Western traditions, such as alchemy and the Kabbalah, as well as Eastern ones, such as taoism. An essential part of these dichotomies between masculine and feminine parts, she maintains, is that they are based upon an idea of dynamic interaction between two essentially equal parts, which is not the case with our present concept of gender.[37] While Singer's approach is also individualistic, and is certainly eclectic, this view at least locates the rigid polarization with which we are familiar in a particular period of history. She does not, however, offer any solutions, other than that the individual attempts to change her/himself.

My problem with these approaches, I must confess, is that I have been brought up in a quite different religion (science), and find it hard to swallow the allusions to mysticism and to a "great age" of matriarchy which characterize such writings. Nevertheless, Singer's advocacy of a notion of "androgyny", a dynamic union between the opposing poles of gender, as a concept which has occurred time and time again in human thought (in myths of creation, for example), is of more than passing interest.

The third possibility is that our tendency to dichotomize is itself a product of biology. That is, dualism results from our "two brains"; the two hemispheres can function as though they are "two brains", dealing with largely different information.[38] I should point out that it is not only the gender dichotomy which is related explicitly to the duality of the brain itself. Popper and Eccles,[39] for example, propose a concept of mind which is transcendent from the properties of the left hemisphere. The right, or "minor" hemisphere, is then seen as influencing consciousness only indirectly. Eccles refers to the right hemisphere as being more "animal". This view, then, retains the mind-body duality and locates it in the dualism of the brain itself. This idea is implied, of course, in the quotations I used above, in describing the associations often made with gender.[40] We are by now tediously familiar with literature describing sex differences in cognitive function and their relation to brain lateralizations,[41] a literature which carries a number of sexist assumptions about gender-appropriate abilities, and which contains several contradictions. The functions of the right hemisphere, for instance, are portrayed by some writers as essentially feminine, while at the same time,

> Men are linked with the right brain because of their supposed superior spatial abilities, women with the left brain because of superior verbal scores. Yet men are also stereotyped as "more analytic" (or "logical") which is said to be a left-brain skill. Complex and circuitous arguments are required to come up with the "superior spatial ability" while leaving the myth of men's razor-sharp intellect intact.[42]

Much, I suppose, depends on the writer's assumptions.

We can, of course, criticize the content of these claims on their own terms, and point out that the data from which inferences of brain laterality are drawn are possibly only valid for own culture (thus casting doubt on the "biological" nature of the data). "Spatial ability", for example, which is held to differentiate the sexes, and which has been related to specific brain areas, varies between cultures, and is not necessarily a universally masculine trait.[43] But that is not the point. The problem with this mode of explanation is that it is biologically determinist, and hence profoundly pessimistic. Having taken a trait which we say is located in such-and-such a part of the brain, and arbitrarily labelled that trait "masculine" or "feminine", it is not far to suggest that becoming masculine or feminine is an essentially *biological* process, to which society only adds the finishing touches. As this way of thinking locates the problems of dualism within the individual, rather than social, domain, there's not much we can do about it, now or in the future.

It seems to me that, in general, these approaches see the tendency to think in dualities as immutable. If the hierarchical gender dichotomies to which feminists object are rooted in a primary, pervasive duality, aren't we all simply wasting our time quibbling over whether these differences are biologically determined or socially conditioned? It certainly seems to occupy our time, and we expend much energy accusing *other* feminists/groups on the left of adopting a position which we eschew (say, accusing someone else of being biologically determinist). [44] Whether we view gender dichotomies as a function simply resulting from a sense of "otherness", whether we refer them to a somewhat ahistorical patriarchal hegemony lasting aeons of time, or whether we view them as residing in the brain, there seems little hope of change. And isn't change what we are trying to achieve?

The views I have mentioned above tend to ignore historical changes. Viewing the world in terms of a gender dichotomy may be an old and nasty habit, but what seems to have changed is the significance we attach to it. The extreme dichotomy now existing between our stereotypes of masculinity and femininity serves a particular function related to the roles men and women are expected to play in our society. "Femininity" and "masculinity" as we now understand them are social constructs which maintain particular relations of power between the sexes, and which themselves have to be continually reinforced – through educational practices, legislation, media representation, and so on. [45] Presumably if they were as biologically based as some writers fondly imagine, we would have little need of this continual ideological reinforcement. The present rigidly separated stereotypes have not existed for millenia: they have grown up in their present form within the context of patriarchal capitalism, which required women and men to fit rather different roles within it. [46] It is in this context that the values attached to masculinity become especially highly valued: it is, therefore, only in this context that we can make much sense of the oppositional constructs which are directly associated with gender.

To recap. There is a tendency for us to view the world as consisting of but two parts, and we commonly associate these parts with gender. This tendency is sometimes seen as a universal. However, many of the suggestions which have been made regarding the origins of this tendency appear to project the values attached to gender by our current society back into other historical epochs. I think this is misleading. Implicit in the ideas to which some of these writers refer is the idea of a dynamic, fluid interaction between two component parts. [47] This is not the same form of dualism as that which dominates thinking in gender dichotomies today. The latter embodies hierarchical relations, which can only be adequately understood in relation to the existing social role of women and men. Total acceptance of these social relations tends to occur, however, within research, as I have indicated in the first section. "Gender" as a rigidly defined construct, becomes an *ideé fixe* running through the hypotheses which are tested, and helps to maintain existing boundaries between behaviour which is acceptable to bourgeois ideology, and that which is not.

A major question which we should ask is what the consequences are of dualistic thought. An immediate effect of dualism is that, when applied to the social world, it creates categories of me and not-me. This need not *necessarily* be a major problem *if* the two categories are equally valued. The problem is that we Westerners have a history of being an arrogant lot, and we are inclined to value those most like ourselves as being better. Although I think that this tendency has got much worse in recent centuries, we were doing it before the rise of capitalism: our arrogance is implicit in the ideas of the Great Chain of Nature, which put humans just a little lower than the angels, and in the Ptolemaic universe, which put *our* home at the centre. But the important question is whether we can envisage a situation in which any dichotomies/categories we may perceive are founded upon equally valued parts. The answer must be no – within the present form of society. That is, a society founded upon profound inequalities and upon individualism. In this sort of society, creating rigid categories can only reinforce existing social divisions.

I have not attempted to address the underlying problem of why we think in dualities in the first place (including Cartesian dualism), but have concentrated on those associated with gender. It may be that, like linearities, it is primarily a product of Western thought, although I doubt it. By concentrating on a particular form of dualistic thinking, I felt it might be easier to see the difficulties and the implications they have for our understanding of reductionism. In considering the problems of reductionism, we have also to consider how we set up the initial categories by which scientific and social questions are defined, and the consequences of this. If we perpetuate the rigid oppositions maintained by society, then we will not be able to see all the possible interconnections. What we have found objectionable about crude reductionism is the tendency to assume a linear causal chain between levels of explanation, such that complex phenomena, such as cognitive processes, or sexuality, become reduced to the level of a few genes, or to a hormone. We also need to recognize that this convergence depends upon the categories by which we define and classify our social world.

Science becomes ideological precisely because it serves to reinforce, rather than to question, those socially defined and prescribed boundaries; and reductionism becomes most politically and philosophically objectionable when it is clearly operating within them. If we are to seek interconnections in order to build a more dialectical biology, then it is necessary for us to question the fundamental categories by which we order our perceptions of the world – whether these be dichotomies based upon gender, or any other classifications. This is a challenging and urgent task.

Notes and references

1. See the papers in R. Needham (ed.), *Right and Left: Essays on Dual Symbolic Classifications*, Chicago 1973.

2. G. Webster, "Scientific Theories and Cultural Classifications: A Comment on Lynda Birke's 'Cleaving the Mind: Speculations on Conceptual Dichotomies' ", unpublished MS.

3. Helen Thompson Woolley, "Psychological literature: A review of the recent literature on the psychology of sex", in *Psychological Bulletin*, no. 7, 1910, pp. 335-42.

4. Sir A. E. Wright, *The Unexpurgated Case Against Woman Suffrage*, London 1913.

5. See, for example, H. Fairweather, "Sex differences in cognition", in *Cognition* no. 4, 1976, pp. 231-80; B. Fried, "Boys will be boys will be boys: the language of sex and gender", in R. Hubbard, M. S. Henifin, & B. Fried (eds.), *Women Look at Biology Looking at Women*, Cambridge, Mass. 1979; D. Griffiths and E. Saraga, "Sex differences and cognitive abilities: a sterile field of enquiry?" in O. Hartnett, G. Boden and M. Fuller (eds.), *Sex Role Stereotyping*, London 1979; and J. Sayers, "Psychological sex differences", in Brighton Women and Science Group (eds.), *Alice Through the Microscope: The Power of Science Over Women's Lives*, London 1980.

6. A well-known hypothesis is that gay men had Over-Intimate Mothers. However, as Davison points out, what is wrong with such a mother unless you happen to find her in the background of someone you have already defined as deviant? See G. C. Davison, "Homosexuality – the ethical challenge", in *J. Consulting and Clinical Psychology*, no. 11, 1976, pp. 163-6.

7. G. Webster, op. cit.

8. B. Ehrenreich and D. English, *For Her Own Good: 150 Years of the Experts' Advice to Women*, New York 1978.

9. J. Weeks, "Movements of Affirmation: Sexual Meanings and Homosexual Identities", paper presented to the Annual Meeting

of the British Sociological Association, University of Sussex, 1978.

10. M. Foucault, *The History Of Sexuality* vol. 1, London 1979; M. McIntosh, "The homosexual role", in *Social Problems* no. 16, 1968, pp. 182-92.

11. See G. Dorner, "Hormones and sexual differentiation of the brain", in *Sex, Hormones and Behavior*, Ciba Foundation Symposia no. 62, 1979, pp. 81-112. See also my critique of the assumptions behind this work: L. I. A. Birke, "Is homosexuality hormonally determined?", paper given to the American Association for the Advancement of Science, January 1979, to be published in N. Koertge and F. Suppe (eds.), *Homosexuality and the Philosophy of Science*, Bloomington 1981.

12. G. Shiveley and J. De Cecco, "Components of sexual identity", in *J. Homosexuality* no. 3, 1977, 41-8.

13. J. Imperato-McGinley, R. E. Peterson, T. Gautier and E. Sturla, "Androgens and the evolution of male gender – identity among male pseudohermaphrodites with 5-alpha reductase deficiency", in *Obstetrics and Gynecology Survey* no. 34, 1979, pp. 769-71.

14. See data in J. Money and A. Ehrhardt, *Man and Woman; Boy and Girl*, Baltimore 1972. The interpretation of these data is, however, open to question: see Lesley Rogers's contribution to this volume.

15. These assumptions are taken over wholesale into animal research. Using animals whose hormone levels have been experimentally changed just after birth, Dorner claims to have created "homosexual" rats. By this he means males which accept the mounts of other males, or females which more readily mount other females. That the control animals also participate in these "sexual" encounters seems to have escaped his notice. See G. Dorner, *Hormones and Brain Differentiation*, Amsterdam 1976.

16. J. W. Ross, L. J. Rogers and H. McCulloch, "Stigma, sex and society: a new look at gender differentiation and sexual variation", in *J. Homosexuality* no. 3, 1978, pp. 315-30.

17. Fried, 1979, op. cit.

18. Money and Ehrhardt, op. cit.

19. Sayers, 1980, op. cit.

20. For example, see S. Bem, "The measurement of psychological androgyny", in *J. Consulting and Clinical Psychology* no. 42, 1974, pp. 155-62; and J. A. Williams, "Psychological androgyny and mental health", in O. Hartnett *et al.* (eds.) op. cit. There are many papers on this theme in the journal *Sex Roles*.

21. For example, L. G. Clemens and B. A. Gladue, "Feminine sexual behaviour in rats enhanced by prenatal inhibition of androgen aromatisation", in *Hormones and Behavior* no. 11, 1978, pp. 190-201; and A. Payne, "Neonatal androgen administration and sexual behaviour: behavioural responses and hormonal responsiveness of female golden hamsters", in *Animal Behaviour* no. 27, 1979, pp. 242-50.

22. R. Hertz, 1909, in R. Needham (ed.), op. cit.

23. S. L. Star, "The politics of right and left: sex differences in hemispheric brain asymmetry", in R. Hubbard *at al.* (eds.), op. cit.

24. R. E. Ornstein, *The Nature of Human Consciousness*, San Francisco 1973.

25. For example, Star, op. cit., and P. Hoch, *White Hero, Black Beast: Racism, Sexism, and the Mask of Masculinity*, London 1979.

26. Hoch, op. cit., and Ehrenreich and English, op. cit., for example.

27. Hoch, op. cit., pp. 141-2.

28. ibid., p. 130.

29. Dr Bean, quoted by I. S. Reid in "Science, politics and race", in *Signs*, no. 1, 1975, pp. 397-422.

30. E.g. Hoch, op. cit., and K. Millett, *Sexual Politics*, London 1971.

31. Ehrenreich and English, op. cit. This association also derives from another way in which we classify the world, that of linear hierarchies. We are here, for instance, discussing a hierarchy of sciences, through which we might infer "downward" causation. Another linear hierarchy has dominated Western thought, and is with us still. I mention it here as it contains within it the gender dualism. The hierarchy to which I refer is the *Scala Naturae*, or Great Chain of Being, an idea which goes back at least as far as Aristotle (A. O. Lovejoy, *The Great Chain of Being: A Study of the History of an Idea*, New York 1936). It involved a conception of the natural world which fitted, more or less, into the Ptolemaic universe (although with some contradictions). According to this view, there existed a linear hierarchy of beings, with God and the angels at the top. Some way below was sinful man, and arrayed below that were animals, plants, and inanimate nature. The Chain was held to be approximately continuous, with every available place filled according to Divine Purpose. Woman's place in this was decidedly below man's. Much debate took place in the Middle Ages about whether women even had souls at all (e.g. Aquinas, in the *Summa Theologica*). People of other races were

slotted into the Chain in due course, but always below Caucasian Man, the height of perfection (the Hottentots, for example, were fitted in between humans and apes). The problem with the Linear Chain is that we still tend to believe in it, Darwin notwithstanding. We refer, for example, to higher or lower animals, or to evolutionary ascent. Dualities emerge from the Chain simply by breaking it at some point. Thus, a division between woman and man emphasizes man's divine associations and woman's earthbound ones. Similarly, we divide "man" off from other species (as in notions of discontinuity), as though we were still operating on a basis of linearity (M. Midgley, *Beast and Man; The Roots of Human Nature*, Hassocks 1978).

32. For example, Sartre: see P. Holland, "Jean-Paul Sartre as a NO to women", in *Sinister Wisdom*, no. 6, 1978, pp. 72-9.

33. R. Aronson, "The individualist social theory of Jean-Paul Sartre", in *Western Marxism: A Critical Reader*, London 1977, p. 228.

34. E. g. M. Daly, *Beyond God the Father*, Boston 1973; E. G. Davis, *The First Sex*, New York 1971; J. Singer, *Androgyny: Towards a New Theory of Sexuality*, London 1977; M. Stone, *The Paradise Papers: The Suppression of Women's Rites*, London 1976

35. See for example H. Hartmann, "The unhappy marriage of marxism and feminism: towards a more progressive union", in *Capital and Class*, no. 8, 1979; and papers in A. Kuhn and A-M. Wolpe (eds.), *Feminism and Materialism*, London 1978.

36. Singer, op. cit.

37. Although this idea of equality of the female and the male principle has characterized certain philosophies, it has not necessarily had any impact on social relations. Imperial China may have been the home of taoism, but it was also the home of foot-binding and widespread female infanticide.

38. See, for instance, M. S. Gazzaniga, *The Bisected Brain*, New York 1970.

39. J. C. Eccles, "Cerebral activity and consciousness", in F. J. Ayala and T. Dobzhansky (eds.), *Studies in the Philosophy of Biology*, Berkeley 1974; and K. Popper and J. C. Eccles, *The Self and Its Brain: An Argument for Interactionism*, London 1977.

40. See discussions in Singer, op. cit., Star, op. cit., and A. Dickason, "The feminine as a universal", in M. Vetterling-Braggin, F. A. Elliston and J. English (eds.), *Feminism and Philosophy*, New Jersey 1977.

41. Many tests of cognitive function which are commonly employed supposedly differentiate the sexes. As a result, several theories

have been proposed to account for these data in terms of processes of biological development (e.g. A. W. H. Buffery and J. A. Gray, "Sex differences in the development of spatial and linguistic skills", in C. Ounsted & D. C. Taylor (eds.), *Gender Differences: Their Ontogeny and Significance*, London 1972). However, much of this is based upon assumptions, rather than upon adequate empirical testing, and supports normative assumptions of masculinity/femininity. And much depends on the dichotomy you use in the first place: the same results from cognitive tests might equally be explained in terms of late maturation/early maturation, rather than sex differences *per se*. For a critical review, see Fairweather, op. cit., 1976.

42. Star, op. cit. p. 70.

43. J. W. Berry, "Ecological and cultural factors in spatial perceptual development", in J. W. Berry and P. R. Dasen (eds.), *Culture and Cognition: Readings in Cross-cultural Psychology*, London 1974.

44. Many of these difficulties, of course, arise from a naive acceptance of the nature/nurture dichotomy: in rejecting crude biological determinism, many feminists implicitly adopt a crude environmentalism. Neither approach helps us much to understand the processes of development. For a more general consideration, see Pat Bateson's contribution to the companion volume to this, *Towards a Liberatory Biology*, London 1981.

45. E.g. M. MacDonald, "Schooling and the reproduction of class and gender relations", in R. Dale, G. Esland, and M. MacDonald (eds.), *Education and the State*, Milton Keynes 1980.

46. See, for instance, Hartmann, op. cit.

47. E.g. Singer, op. cit.

5
The Ideology of Medicine
Lesley Rogers

Reductionist thinking has always been the ideology of Western medicine, as it has of Western science. Diseases are reduced to microbial causes, microbes to molecules, people to bodies with molecular functioning, and so on. At one level, this mode of thinking has been enormously successful; for example, in understanding the physiological make-up of the body and finding the cures for many physical diseases. However, there is debate about what we can call a purely "physical" disease, and when reductionist thinking is applied to diseases which encompass more than biological functioning, to the extent of interacting with behaviour of the individual and the societal structure of which the individual is a part, it is totally inadequate. It is not at all difficult to see that this is clearly the case in medicine's attempts to explain "mental disease",[1] but it is also true for a number of diseases once thought to be purely physiological in origin and cure (e.g. heart disease, cancer, those which the medical profession calls "diseases of malnutrition", such as Kwashiorkor, and even conditions like gallstones, respiratory infections and Parkinson's Disease[2]). Medicine is becoming increasingly aware that people heal better in some social environments than they do in others, and that a poor social environment of one sort or another can lead to mental plus physical illness even to the point of death. These days there is frequent talk of "the sick society" or "the illness in society", but even though these terms imply that there is something wrong with society, the blame for "the sick society" is still placed upon individuals and not the society itself. For example, the rise in violent crime is generally believed to result from some strange disease infecting a large portion of the lower classes rather than an expected outcome of class-stratified capitalist society.

Even though medicine has made some recognition of the importance of factors outside molecules and physiology, this recognition has not as yet brought many changes in the practice of medicine, except to a limited extent in isolated cases where doctors are practising what is called "community medicine". Community medicine includes preventative medicine, self-help groups, more follow-up treatment, etc.

But the hospital itself, public or private, retains essentially the same organization that it developed in the days of Florence Nightingale. While it is quite generally accepted that patients recover from operations faster when they are in public wards rather than private wards, private wards still exist and the rich are not discouraged from using them. It is by looking at the practice of medicine that we gain a great deal of insight into its dominant ideology.

Hospitals are run like military organizations, based on class divisions, race divisions, divisions between the sick and the well, between the so-called "mentally" sick and physically sick, and between men and women. According to these categories people are placed in different wards, often different hospitals, and receive different treatment; that is, if they receive treatment at all. Many blacks in Australia receive no Western medical treatment, and those who do almost invariably receive an inferior quality. Also in Australia, the aged pensioners and recent immigrants (mostly Greeks and Italians) are treated in the large public hospitals and must frequently wait for years for some operations which would be done immediately if they could pay for private hospital treatment. For problems related to the reproductive organs, or the sex-role differences between men and women, women are more likely than men to receive drug therapy or surgery and less likely to receive time-consuming counselling (e.g. in Australia, at least 40 per cent of women receive hysterectomies by the time they have reached the age of 65, and approximately half of these are performed for menstrual symptoms and in the absence of physical pathology).

The stratified practice of medicine is coupled with hierarchical structure within the profession itself. The holy orders of surgeon, registrar, house doctor, sister, nurse, orderly and so on, are rigidly defined and maintained, this stratification being a facet of the capitalist society in which it exists. (Compare the organization of medicine in, for example, China.) This stratified medical practice is enormously successful in extracting money from the sick and the poor. Not only is it an aspect of capitalism, but a very powerful and significant force in perpetuating capitalism. This is true at the level of its practice, and even more so at the level of its ideology. Popular thought and theories in medicine have great power in controlling the attitudes of society. For this reason, I am most concerned about a recent upsurge of interest in reductionist explanations for human behaviour. There has been thankfully, a reasonably large amount of criticism of sociobiology, but meanwhile the reductionist trend in medicine, which has much in common with sociobiology, has been gaining strength with relatively little criticism.

It is not surprising that when psychiatry emerged it carried with it reductionist ideology. "Mental disease" was, and still is, considered to be caused by a derangement of the molecular functioning or structure of the brain, which, when discovered, will lead to cures by the appropriate drug therapy or psychosurgery. Indeed some psychiatrists act under the delusion that these things are already proven (e.g. psychosurgery for sexual "deviants"[3]).

Behaviours like the ones labelled schizophrenia, mania, depression, homosexuality were thought to be genetically determined, and geneticists have sought to substantiate this.[4] Criticism of these studies similar to that applied to the studies of Eysenck and Jensen can be raised here. Suffice it to say that the genetic basis of these conditions has not been proven, but the theories are still ardently held by a large section of the medical profession and the community at large.[5]

With the emergence of psychopharmacology in the early 1950s, new hope of finding molecular cures for "mental disease" sprang up.[6] Use of "wonder drugs" spread rapidly around the world in a wave of drug-manipulation of people which is still gathering strength today, long after it has been realized that these drugs cannot cure, but merely serve to suppress the symptoms (i.e. dull individuals into a state where they no longer react).[7] Despite the lack of success of psychopharmacology, the initial belief systems have remained unshaken; psychopharmacology is an expanding field still promising to find, eventually, the cause and cure for mental illnesses.[8]

It is worth diverging temporarily at this point to discuss what I see as a house-of-cards situation in psychopharmacological therapy. Almost all psychopharmacological agents have been discovered by accident rather than by design. For instance, chlorpromazine was being tested as an antihistaminic when it was discovered to have a strong effect on behaviour, and thus it was soon administered to psychotic patients, by Delay in 1952. The anti-depressant, iproniazid, was being used to treat tuberculosis when it was noticed to have mood-elevating effects. Other psycho-active drugs, like reserpine, come from folk medicine in various cultures. Chlorpromazine and reserpine are both anti-psychotics but their structures are vastly different. This is the case for all psychoactive drugs, and it is therefore impossible to categorize them primarily according to structure. They are classified into their primary categories according to the behavioural conditions on which they are noticed to have most effect and not according to their chemical structures. Sub-categories are based on chemical structure, since many new drugs have been developed by making alterations in the basic structures of those first found to be effective by trial and error. However, relatively slight changes in chemical structures can radically alter the action of CNS drugs, thus requiring that they now be classified into a new primary category. Hence another reason for the need to categorize primarily on the basis of clinical practice.

Now if we pick up a text-book on psychiatry, we will see that the initial diagnosis is the most important step in psychiatric treatment[9] and we will soon become aware that the divisions between the various categories of "mental illness" are not clear cut.[10] Therefore one tool used in diagnosis is to see which of the psycho-active drugs will remove the symptoms![11]

Thus here we have an example of reductionist thinking operating on the one hand in psychopharmacology, with a drug classification relying on what it believes to be the hard and correct classification of mental illness by psychiatry, and on the other hand psychiatric classification relying on what it believes to be the securely known clinical actions of drugs. Even within two mutually dependent areas of medicine reductionist thinking has narrowed the sights so that what at first appears to be a mutual support structure of ideas and practice collapses when examined just a little more broadly. Yet despite this situation, and despite the lack of success researchers have had in finding genetic or molecular causes for "mental illness" or "sexual deviation", interest is increasing and, particularly in the U.S.A., enormous amounts of money are being spent on research in this area. Many medical scientists get this money by claiming they are on the brink of discovering the cause of schizophrenia.[12] One should emphasize that this sort of research takes no account of the social factors surrounding the person who has been labelled schizophrenic, or the society in which the disease occurs.

Causal theories, such as schizophrenia being caused by excess dopaminergic activity in the brain[13] and "hyperactivity" in children being caused by a dopamine deficiency,[14] are deduced from the assumed action of the drugs used to suppress the symptoms of these "diseases" (e.g. antipsychotic drugs antagonize the action of dopamine, plus other neurotransmitters, and amphetamine which is given to "hyperactive" children is thought to activate dopaminergic pathways). Some scientists have expressed the hope that, when such causal theories are proven, we will be able to diagnose at least a sub-set of these "diseases" by making biological measurements, such as assaying for metabolites for dopamine in body fluids (e.g. homovanillic acid in cerebrospinal fluid).[15] Such thinking reduced what was once considered to be a behavioural problem to a biological one. Indeed, it has been suggested that we may be able to detect potential schizophrenics and hyperactive children by such biological analysis,[16] and act accordingly!

The anti-psychiatry school has struck out against drug therapy and opted for social causes of madness, seeing it as a normal response to an abnormal situation.[17] And so we have the nature and nurture schools of psychiatry. Although there is more convincing evidence in favour of nurture over the nature explanation, I do not see that either explanation is sufficient. While such schools continue with virtually no regard for each other, perhaps the real explanation lies not somewhere in between, as many fence-sitters have exclaimed, but somewhere else. Of course, one can immediately say that we need to take an approach which allows for the interaction of biology and environment (society) and that prevents the separation of mind from body. I am in favour of that, but exactly how do we do that and how do we research it? Sequential explanations, like those which say that the family or society causes a person to behave as a schizophrenic and this causes the biology to alter in turn, or alternatively that some people are *predisposed* to be schizophrenic and the environment interacts either to prevent it or to precipitate it, do allow for the contribution of both environment and genes, but there are also problems with this approach. We see a similar argument put forward by Money and Dalery[18] in their biological explanation for homosexuality. They say that young girls can be predisposed to being lesbian by being exposed to excess androgen levels before birth, and their upbringing after birth either suppresses or enhances this. For example, they propose "a formula for creating the perfect female homosexual";[19] take an individual with female genetic code, expose her to androgen in the foetal stages of development thus producing masculinization of the genitalia, and follow this by rearing her as a boy! Any relationship this has to homosexuality is obscure, if not negligible,[20] but I will not pursue this aspect of the issue. I use it simply as an example of the kind of medical thinking that claims to allow for the interaction of nature and nurture, but fails to do this, since at any one point in time nature takes precedence over nurture, or vice versa. Such approaches are often ludicrously oversimplified and dangerous.

The idea of "predisposition" is, I believe, becoming increasingly common. At the last International Ethological Conference (in Vancouver, 1979) I noticed that it is a position into which many sociobiologists are now retreating. If a characteristic is not entirely genetically determined, then let us say that the genes set a "predisposition" for it. That way the theory is less easily pinned down and criticized, but in terms of its societal or political impact it is just about as effective as is the straightforward genetic determinist position.

Genes and environment interact, and they do so at all times in such a way that they cannot be separated. It is not possible, therefore, to say that genes set a predisposition before birth and environment plays the main role after. This latter approach can, I think, be considered as "sequential reductionism", rather than interaction.

During the 1960s and 70s there has been a developing interest in the psychosexual sect of medicine, and John Money has been the central figure here. When medical schools (in Australia, at least) began introducing courses on "Human Sexuality", John Money's prolific writings (both in the psycho-medical literature and in more popular publications) became their main reference. The attempt was to explain sex differences in behaviour, homosexuality, transvestism, trans-sexualism and the so-called "premenstrual syndrome", on the basis of biology. John Money and his co-workers claimed to show that females exposed to androgen before birth (either because they had the Adrenogenital Syndrome, a congenital condition in which the adrenal glands secrete excessive amounts of androgens, or because their mothers had been given progestins – which sometimes have an androgen-like effect – to prevent miscarriage) had a higher IQ, preferred career over marriage, preferred boy's clothes, were late to reach "the romantic age of dating" and had confused gender identity.[21] I have criticized the sexist bias of this work and its bad methodology in detail before.[22]

In all the studies of Money and his co-workers there are implicit assumptions that the categories above denote gender-related behaviours which have been shown to differentiate consistently between "normal" boys and girls, while these have by no means been substantiated.[23] The same is true for more recent studies by Ehrhardt and Baker[24] in which Maccoby and Jacklin[25] are quoted to substantiate these assumptions. No recognition has been given to sex role socialization research,[26] which shows that it becomes far more difficult to isolate gender-related behaviour in which boys and girls typically differ when sex role socialization patterns are changing as they have been in the seventies. Indeed, throughout this psycho-medical area of research there has been a total lack of consideration of sociological factors, which are so obviously important to the results, and their interpretation (e.g. the attitudes of the parents to their androgen-exposed daughters, and the influence of social class which differed between the control and experimental groups: see note 22). If you are going to push forward a biologically determinist theory for explaining sex differences in behaviour, I guess this sort of ignorance of contributing variables helps. Bad methodology would also seem to help; for example, "tomboyishness" was scored by asking the mothers whether their daughters liked energetic play (climbing trees, etc.). It is not difficult to show the methodological flaws in this work, but here I wish to concentrate on their interpretations and general theories.

Money and Ehrhardt suggest that their findings may indicate links between IQ, "tomboyishness", and choice of career over marriage,[27] and that dominance behaviour and academic achievement may be "affected positively in girls exposed to an excess of pre-natal androgens, while maternalism is negatively affected".[28] The underlying causal explanation put forward for these differences between men and women is a biological one; extrapolating from their claim that higher IQ occurs as a result of androgen exposure during development, they raise the possibility that the development of the cortex and its concomitant intellectual function is affected by the sex hormones. If a programme of research were conducted to test this hypothesis, it would not be a far cry from the old psychological studies which attempted to demonstrate a relationship between cranium size and intelligence by immersing black and white heads in buckets of water, or from studies in the late 1800s which attempted to verify the believed mental inferiority of women by studying size, weight, shape and fissure patterns in brains.[29]

In their follow-up study Baker and Ehrhardt[30] failed to confirm that high IQ and Adrenogenital Syndrome were related, since unaffected members of the subjects' families had equally high IQ scores, and no Adrenogenital Syndrome. Instead of taking this as evidence against the original theory for biological determinism of IQ, they went on to propose that there may be a recessive gene which causes Adrenogenital Syndrome which is linked to another dominant gene for higher IQ.

In the theories put forward by Money et al., reductionist thinking is inaccurately applied even within the biological level; the differentiation and function of the brain is reduced to that of the sex organs. The brain is assumed to differentiate into either a male or female form with a concordant male or female behavioural function under the influence of androgens in exactly the same way that the gonads are known to differentiate into the male or female form. At the cellular level this theorizing equates the infinitely complex structural organization of the brain with the much simpler structure of the gonads; more seriously, at the functional level the performance of the brain and its translation into behaviour of the whole individual in society is equated to structural differences between male and female gonads. There is no evidence to support this grossly simplistic thinking.[31]

It is important to mention that there are times when Money and his co-workers[32] give definite credence to the cultural contribution to sex-role (e.g. the case of *penis ablatio* of an identical twin subsequently raised as a girl and reported to be entirely "feminine in behaviour").[33] However, the cultural influence is never fully integrated with the biological. There is a very obvious contradiction between the practical side of their research and their theoretical discussion of gender identity, homosexuality, etc. In fact there is often a direct contradiction between the theory they propose and claim to have evidence to prove on one page and the data quoted several pages later.[34] These contradictions are couched in so much medical jargon and convoluted reasoning that the reader is swept along bewildered and is hoodwinked into thinking that an integrated theory involving biology and culture is being put forward. Many readers have been taken in by this, and have been left with the feeling that they did not quite understand it all, the ignorance being theirs, not Money's and his co-workers'!

Money and co-workers adhere to the idea of "predispositions"; sex hormones have a causal influence on brain differentiation in early life and cultural learning can modify this later. For example:

> What your prenatal mix did was to lower the threshold so that it takes less of a push to switch you on to some behaviour and to raise the threshold so that it takes more of a push to switch you in to other kinds. More androgen prenatally means that it takes *less* stimulus to evoke your response as far as strenuous physical activity or challenging your peers is concerned, and *more* stimulus to evoke your response to the helpless young, than would otherwise be the case.[35]

This sequential explanation has its roots firmly planted in biological determinism, as can also be seen in Money's discussion with respect to "tomboyishness", preference in clothing, career ambition, gender identity, etc:

> I think these components are, in origin, pretty much independent of rearing, even though they are influenced by the negative and positive

reinforcements that a child gets from parents and others in the course of rearing.[36]

While the only research that Money and his co-workers have done has been using subjects exposed to extreme levels of androgens before birth, they extend these findings to explain the gender-related behaviours of all males and females. Even if they had produced genuine results, which they did not, these clinical examples may have no bearing on male-female differences in general.

By providing a rigid biological basis for sex differences in behaviour, such theories subserve the oppression of women. There have been others who have adopted a similar line. For example Hutt,[37] Buffery and Gray,[38] and Lambert[39] explain sex differences in behaviour on the basis of *either* genes *or* hormones affecting the degree of lateralization in the brain (see Lowe[40] for discussion of these). And we should not forget E. O. Wilson[41] and his belief that present day sex-roles are no different from those of hunter-gatherer societies, which in turn are no different from animal societies, and so they have been genetically determined over the course of evolution.

The *raison d'être* of all of these theories is to maintain the present inferior position of women in society. A secondary effect they have is to provide biologically based explanations for the so-called "sexually deviant" groups. Money and Ehrhardt,[42] like several others (e.g. Lorraine[43] and Wilson *et al.*[44]), adhere to the idea that "sexual deviation" is caused by errors in development. These errors are either genetic in origin (and according to Lorraine there are more male homosexuals than female because the XY genotype allows for expression of more genetic errors), or they may result from a discordance between early androgen exposure and post-natal learning.[45] Money[46] places homosexuality, transvestism and transsexualism along a continuum of increasing disturbance of androgen levels; he is, of course, unable to give hormonal evidence to substantiate this (see Ross *et al.*[47]). None of the androgen-exposed females in his study were either transsexual or homosexual.[48]

Money certainly sees himself as taking a progressive and liberal approach to sexuality, and regrettably many new courses on "Human Sexuality" which are being introduced in Australian medical studies accept this view. Money's writings and related beliefs often form the basis of these courses. While Money may appear "liberal" in that he discusses topics hitherto taboo, and argues for some narrowing of the gap between male and female roles, no person who writes that sex roles "are the glue which holds society together"[49] is anything more than a conservative force working against the feminist and gay movements. No person who wrote,

> If the stereotypes are too amorphous, the society fails to provide its members with the necessary means of co-operation and soon falls apart. The tendency of cultural stereotypes to resist change is essential for maintaining society,[50]

at the time when the feminist movement was questioning the usefulness of sex-roles, and when there was a strong and growing movement towards removing them altogether, can be considered a liberal progressive. Seen in context, Money was, and still is, part of the backlash thrown up by society to curtail the movement for changing sex-roles.

The field of medicine which concerns itself with these issues is a pseudoscience with enormous political implications. Yet in contrast with sociobiology and the IQ racket that preceded it, psycho-sexual medicine has received surprisingly little criticism. I came across one reason for this when Walsh and I wrote a critical paper[51] and then sought for an appropriate journal in which to publish it. We discovered that every one of the dozen or so medical journals held in our university's library and which publishes this sort of material, has on its editorial board either John Money or one of his collaborators. We were thus unable to express a contrary view in the journals in which Money publishes so prolifically, but were forced to go to sociological journals, not even housed in our Medical School library. Thus while an expanding and already discrete field has formed, it is still small enough to be dominated by a handful of people. Of course it is possible to attack these ideas in publications not overlapping with the ground staked out by Money et al., and there have been some criticisms here and there,[52] but the opposition has had nothing like the power and fervour of that against the IQ story and sociobiology. I think this is because the latter is being aimed (very admirably) against racism, while theories and practices which oppress women, homosexuals, etc. are still seen to be less important. In the case of women, at least, I guess they are too close to home.

Finally, one may ask what kind of studies can be done in these and related areas. First, I can stress the need for an eclectic approach which considers in depth not only biological factors but psychological, social and political factors in one and the same study. Consider cyclic changes in behaviour in women across the menstrual cycle. The medical profession has said that these are caused by hormonal changes and that there is a real set of symptoms, termed "premenstrual syndrome", which is hormonally caused and should be used to keep women out of equal employment with men: "You wouldn't want the president of your bank making a loan under the raging influence of that particular period" (E. Berman, M.D., *New York Times*, 26 July 1970). This is also used to explain the believed inferiority of women, as can be seen from the following quote from the introduction to K. Dalton's latest book on menstruation:

> It has also been written to help men to understand the capricious and temperamental changes of women, so that the image of women as uncertain, fickle, changeable, moody and hard-to-please may go, to be replaced with the recognition that all these features can be understood in terms of the ever-changing ebb and flow of her menstrual hormones.[53]

There have been numerous medical studies with methodology similar to that of Money and Ehrhardt, in which correlations between behaviour and hormone levels across the cycle have been interpreted as causally connected, from hormones to behaviour. At the same time, there are numerous studies in the sociological literature which consider only the social factors which may contribute to this, and no biology. No correct understanding of this or related areas will be available until studies are done in which biological and sociological factors are considered in one and the same study.[54] A broader approach using better methods to collect data (e.g. not just questionnaires, but, say, an ethnomethodological approach; see note 55) will provide us with better information. The data must then be correctly interpreted. Here, I think, is where we need new ideas, and unfortunately I cannot provide them. All I can say is that correlations must not be seen as causations,[56] and authors of papers must be certain to make strong statements which will work against the traditionally reductionist framework into which the minds of most of their readers will imperceptibly slide. This I see as their social responsibility both in educating their readers, and in working against those forces in society which will seize on any opportunity to gather data which can be twisted into giving the appearance of scientific evidence in support of our corrupt and oppressive society.*

---

* The views expressed in this paper are not necessarily shared by the other members of the Pharmacology Department, Monash University.

Notes and references

1. The problems of applying the medical model to explain so-called "abnormal behaviour" have been discussed elsewhere. See, for instance, E. M. Bates, "Alternative theories of madness", in E. M. Bates and P. R. Wilson (eds.), *Mental Disorder or Madness?*, Queensland 1979.

2. O. W. Sachs, *Awakenings*, Harmondsworth 1976.

3. I. Rieber and V. Sigush, "Psychosurgery on sex offenders and sexual 'deviants' in West Germany", in *Archives Sex. Behav.*, no. 8, 1979, pp. 523-27.

4. P. Mittler, *The Study of Twins*, Harmondsworth 1971, pp. 119-45; A. Bertelsen, B. Harvald and M. Hauge, "Danish twin study of manic depressive disorders", in *Br. J. Psychiatry*, no. 130, 1977, pp. 330-51.

5. W. G. Clark and J. del Guidice, *Principles of Psychopharmacology*, London 1978.

6. V. G. Longo, *Neuropharmacology and Behaviour*, San Francisco 1972.

7. See M. Lader, "Benzodiazepenes – the opium of the masses?", in *Neuroscience*, no. 3, 1978, pp. 159-65. There may be a valid argument for short-term use of psychoactive drugs, to assist a person, say, into a state in which they can communicate. But this is only for days, or at most weeks, and certainly not years; whereas most people who were put on to anti-psychotic drugs in the 1950s are still taking them thirty years later.

8. A. Carlsson, "The impact of catecholamine research on medical science and practice", in E. Usdin, I. J. Kopin and J. Barchas (eds.), *Catecholamines: Basic and Clinical Frontiers*, New York 1979, vol. 1, pp. 4-19.

9. F. A. Whitlock, "The traditional psychiatric view", in E. M. Bates and P. R. Wilson (eds.), *Mental Disorder or Madness?*, Queensland 1979.

10. A. A. Buss, *Psychopathology*, New York 1976.

11. Hence we frequently find people in mental hospitals in Australia being given several different kinds of psychoactive drugs at once (e.g. major tranquilliser + lithium + antidepressant). This means that drug therapy has been commenced without initial diagnosis, and the approach is later to discontinue one or two of the drugs to see what happens (i.e. to use the drugs to diagnose: see also G. Bignami's contribution to this volume). Certainly the effects of the drug mixture would prevent diagnosis by behaviour *per se*.

12. M. Clark, M. M. Gosnell and D. Shapiro, "Drugs and psychiatry: a new era", in *Newsweek*, 12 November 1979, pp. 52-7.

13. G. J. Siegel, R. W. Albers, R. Katzman and B. W. Agranoff, *Basic Neurochemistry*, Boston 1976, pp. 731-33; A. Carlsson, "Mechanisms of action of neuroleptic drugs", in M. A. Lipton, A. DiMascio, and K. F. Killamn (eds.), *Psychopharmacology: A Generation of Progress*, New York 1978, p. 1068.

14. M. Clark, D. Shapiro, M. Hager, J. Huck and P. Abramson, "The curse of hyperactivity", in *Newsweek*, 23 June 1980.

15. J. D. Barchas, H. Akil, G. R. Elliot, R. B. Holman and S. J. Watson, "Behavioural neurochemistry: neuroregulators and behavioural states", in *Science*, no. 200, 1978, pp. 964-73.

16. ibid., and Clark *et al.*, op. cit. (note 14), 1980.

17. R. D. Laing, *The Politics of Experience*, Harmondsworth 1973; T. S. Szasz, *The Myth of Mental Illness*, London 1972.

18. J. Money and J. Dalery, "Iatrogenic homosexuality", in *J. Homosexuality*, no. 1 (4), 1976, pp. 357-71.

19. ibid.

20. The homosexual role is not related to gender identity or to "masculinity"/"femininity". Lesbians are not more "masculine" than other women, nor do they wish to be male. Similar arguments apply to male homosexuals. See M. Ross, "Relationship between sex role and sex orientation in homosexual men", in *New Zealand Psychologist*, no. 4, 1975, pp. 25-9.

21. J. Money and A. A. Ehrhardt, "Pre-natal hormonal exposure: possible effects on behaviour in man", in R. P. Michael (ed.), *Endocrinology and Human Behaviour*, Oxford 1968; J. Money and A. A. Ehrhardt, *Man and Woman: Boy and Girl*, Baltimore 1972.

22. See L. J. Rogers, "Biology and human behaviour", in J. Mercer (ed.), *The Other Half: Women in Australian Society*, Harmondsworth 1975; L. J. Rogers, "Male hormones and behaviour", in B. Lloyd and J. Archer (eds.), *Exploring Sex Differences*, London 1976; L. Rogers, L. and J. Walsh, "Shortcomings of psychomedical research into sex differences in behaviour: social and political implications", in *Sex Roles* (in press, 1980); M. W. Ross, L. J. Rogers and H. McCulloch, "Stigma, sex and society: a new look at gender differentiation and sexual variation", in *J. Homosexuality*, no. 3 (4), 1978, pp. 315-30.

23. See, for example, H. Fairweather, "Sex differences in cognition", in *Cognition*, no. 4, 1976, pp. 231-80.

24. A. A. Ehrhardt and S. W. Baker, "Fetal androgens, human central nervous system differentiation, and behavior sex differences", in R. C. Friedman, R. M. Richart and R. L. Van de Weile (eds.), *Sex Differences in Behaviour*, New York 1974, pp. 33-51.

25. E. E. Maccoby and C. N. Jacklin, *The Psychology of Sex Differences*, Stanford 1974.

26. For example, R. Hartley, "A developmental view of female sexrole identification", in B. J. Biddle and E. J. Thomas (eds.), *Role Theory: Concepts and Research*, New York 1966; C. Joffe, "As the twig is bent", in J. Stacey, S. Bereaud and J. Daniels (eds.), *And Jill Came Tumbling After: Sexism in American Education*, New York 1974.

27. J. Money, J. and A. A. Ehrhardt, *Man and Woman: Boy and Girl*, Baltimore 1972, p. 102.

28. ibid, p. 245.

29. F. Salzman, "Are sex roles biologically determined?", in *Science for the People*, July/Aug. 1977, pp. 27-43.

30. S. W. Baker and A. A. Ehrhardt, "Pre-natal androgen, intelligence, and cognitive sex differences", in R. C. Friedman, *et al.* (eds.), *Sex Differences in Behaviour*, New York 1974, pp. 53-84.

31. See, for example, F. A. Beach, "Hormonal factors controlling the differentiation, development and display of copulatory behavior in the ramstergig and related species", in E. Tobach, L. R. Aronson and E. Shaw (eds.), *The Biopsychology of Development*, New York 1971.

32. Money and Ehrhardt, op. cit. 1972 (note 27).

33. J. Money, "Ablatio penis: normal male infant sex-reassignment as a girl", in *Arch. Sex. Behav.* no. 4 (1), 1975, pp. 65-71.

34. Money and Ehrhardt, op. cit. 1972 (note 27).

35. J. Money and P. Tucker, *Sexual Signatures: On Being a Man or Woman*, Boston 1975.

36. J. Money, "Sexually dimorphic behavior, fetal hormones, and human hermaphroditic syndromes with a note on XYY", in H. C. Mack and A. I. Sherman (eds.), *The Neuroendocrinology of Human Reproduction*, Proc. 4th Annual Symposium on the Physiol. and Pathol. of Human Reproduction, 1971, pp. 183-9.

37. C. Hutt, *Males and Females*, Harmondsworth 1972.

38. A. W. H. Buffery and J. A. Gray, "Sex differences in the development of spatial and linguistic skills", in C. Ounsted and

D. C. Taylor (eds.), *Gender Differences: Their Ontogeny and Significance*, London 1972.

39. H. H. Lambert, "Biology and equality; a perspective on sex differences", in *Signs*, no. 4, 1978, pp. 97-118.

40. M. Lowe, "Sociobiology and sex differences", in *Signs*, no. 4, 1978, pp. 118-25.

41. E. O. Wilson, *Sociobiology: The New Synthesis*, Cambridge, Mass. 1975.

42. Money and Ehrhardt, op. cit. 1972 (note 27).

43. J. A. Lorraine (ed.), *Understanding Homosexuality, its Biological and Psychological Bases*, London 1974.

44. G. Wilson and D. Wias, *Love's Mysteries: The Psychology of Sexual Attraction*, London 1976.

45. Money, op. cit. 1971 (note 36).

46. J. Money, "Two names, two wardrobes, two personalities", in *J. Homosexuality*, no. 1, 1974, pp. 65-70.

47. Ross, Rogers and McCulloch, op. cit. 1978 (note 22).

48. Note that the use of these labels as discrete entities is in itself part of the process of reductionist thinking (see Lynda Birke's contribution to this volume). The existence of these labels is not paralleled by equivalent, discrete categories in the real world. Indeed, use of these labels is part of the process of defining a problem which then must be explained or "cured".

49. Money and Tucker, op. cit. 1975 (note 35).

50. ibid., p. 10.

51. Rogers and Walsh, op. cit. 1980 (note 22).

52. Salzman, op. cit. 1977 (note 29).

53. K. Dalton, *Once a Month*, Hassocks 1979.

54. L. J. Rogers, "Menstruation", in *Aust. Family Physician*, no. 8, 1979, pp. 923-39.

55. S. J. Kessler and W. McKenna, *Gender: An Ethnomethodological Approach*, New York 1978.

56. H. Rose and S. Rose, *The Political Economy of Science*, London 1976, Chapters 6 and 7.

6

Disease Models and
Reductionist Thinking
in the Biomedical Sciences

Giorgio Bignami

Introduction[1]

Reductionist models, as has been underlined by M. Barker in his chapter,[2] must be viewed primarily as ideological constructions. In fact, an important step in their analysis and demystification is to understand when, where, how, and by whom, "commonsense" concepts and explanations serving identifiable purposes (political, socio-economic, etc.) have been placed in the hands of specialists to give them sanction by appropriate hypotheses or theories couched in scientific terms.[3]

The biomedical universe certainly constitutes an inexhaustible source of examples of such a process. In fact, quite early in the history of humankind, medical treatments and medical models (or their equivalents) became part of highly effective control systems that served to buffer the more dangerous contradictions and to provide acceptable explanations for both un-ease and dis-ease states.

Since the rise of the positivistic method in the nineteenth century this process has acquired a reifying quality based on the use of reductionist pathological models, which aim at closer and closer analogies between the expropriated human body and other important components of the capitalistic structure.[4] So the questions "who is sick, and why, and how? What does illness mean to the diseased subject?" have gradually been turned into questions such as "which part of the body is worn out or broken, and why, and how can it be repaired or replaced? And in case repair or replacement is impossible, can we scrap the useless part without losing all useful function, or should we scrap the whole thing?"

As Achard and co-workers have convincingly shown,[5] the original reifying tendencies in the reductionist models have gradually led to an

94

impressive semantic and logic structure which embraces a very wide area, ranging from physics, chemistry and molecular biology to welfare administration, sociology and politics.

The same authors also show that at the surface the fundamental disciplines – particularly basic biology founded on mathematics, physics and chemistry – appear now to dictate the rules of the game, providing a "truly scientific" basis for medicine. In the substance, however, it is almost always medicine – i.e. the discipline that must face most directly the contradictions arising from un-ease and dis-ease – which identifies the direction in which practical solutions must be sought, and which proposes raw versions of explanatory models. One step further, and medicine proceeds to enslave the fundamental disciplines which must find for such solutions and models adequate scientific bases.

Last, but not least, the dictatorial role of medicine extends not only "downwards", towards the basic sciences, but also "upwards", towards the behavioural, the social and the political sciences. In fact the long experience of physicians in coping with un-ease and dis-ease, their ability to survive as a compact body of technicians in spite of repeated potentially disastrous failures, their shrewdness in ascribing to their discipline many important changes due to non-medical causes,[6] plus their success in enslaving the fundamental disciplines, have recently led to the adoption of a similar strategy in areas in which different types of mystifications had so far prevailed. The turning of psychology into psychobiology, of sociology into sociobiology, and of politics into biopolitics are, *mutatis mutandis*, extensions of the strategy leading to medicalization of human problems.

It must be obvious at this point that a subject as wide as that outlined in this introduction should not be squeezed within the limits of the present contribution; therefore, from this point on, I will proceed to illustrate some poorly known aspects of the construction, promotion, and internalization of reductionist biomedical models.

The Ex Juvantibus *logic; diagnostic and aetiopathogenetic inferences*

By definition, any physician with an adequate professional standard (relative to the time and place of his or her work) must first attempt curative interventions, and then, if cure is impossible, provide the best available palliation. This has always included, and still includes, the use of intrinsically ineffective therapeutical tools to provide psychological support, the importance of which cannot be minimized – what yesterday went under the expression *ut aliquid fieri videatur* ("so that it can appear that something is being done") and today belongs to the ever-growing area of *placebo* effects.

For a fairly long time the single most important element in this process has been not so much the greater or lesser effectiveness of the intervention *per se*, but rather the collective appreciation – by the physician and patient and/or other subjects with whom the patient entertains privileged relations – that the course of the ailment has been in some way *modified* by the intervention itself. (Later on it will be shown that in certain instances not only modifications for the better, but also modifications for the worse can serve the purpose). Consequently, it is not surprising that medical explanations should often use as starting points possible mechanisms of action of those treatments which have, or appear to have, modified the course of a given ailment.

In fact, it is far from unusual that clinicians, either alone or in good company with representatives of those research areas which are more readily abused by medicine, agree on statements, or working hypotheses, belonging to one of the following types.

Type 1, first example: Patient A, but not patient B (both clinically depressed) have improved during treatment with drug X (a tricyclic antidepressant); therefore, it appears likely that A's depression is endogenous and B's depression is reactive. *Second example:* If a severely anxious patient fails to show improvement under a conventional anti-anxiety treatment, but responds well to a neuroleptic (i.e. to an "antipsychotic" agent), there may be a psychotic, rather than a neurotic state underlying the anxiety syndrome.

Type 2, first example: Clinically effective antidepressants modify monoamine metabolism via a blockade of reuptake at terminals, which enhances the availability of given neurotransmitters; therefore, depression may be the consequence of reduced neurotransmitter contents or availability. *Second example:* Effective antipsychotic (neuroleptic) agents are powerful dopamine antagonists; therefore, the cause (or primary pathogenetic mechanism) of schizophrenia may reside in a disturbance of dopamine metabolism.

Because of the special significance of reductionist models as explanations of behaviour, these examples, drawn from biological psychiatry,[7] show how apparently desirable consequences of a given treatment procedure are often used for two purposes besides cure or palliation. The first is confirmation of a diagnosis (or an operation called differential diagnosis) in a situation in which the observed symptoms lend themselves to more than one type of inference. The second consists of moving "backwards from what helps" – hence the terms *a posteriori* or *ex juvantibus* – to the aetiology and the pathogenesis of a disease state, based on the assumed mechanism of action of the treatment employed.

While there is no doubt that the *ex juvantibus* logic can be useful in a variety of circumstances (see later) it appears obvious that in the absence of adequate independent evidence from other sources it can easily lead to wrong conclusions at all levels. In the cases quoted above, for example, differential diagnoses made on the basis of differential treatment effects are, at best, shaky constructions; this applies both to endogenous versus reactive depression and to psychotic versus neurotic anxiety.

Furthermore, the disease models inferred on the basis of drug mechanisms of action are grossly inadequate from at least two viewpoints. First, they ignore the fact that complex behavioural problems require consideration of historical, socioeconomic and cultural factors, and cannot be adequately described in pathophysiological-neurochemical terms which do not provide any account, e.g. of important cognitive aspects. This ignorance leads to a reduction of the problems themselves to a simplistic biochemical model which can, at best, account for a particular aspect of the changes which may have occurred. Secondly, the models tend to confound hypothetical "downstream" mechanisms contributing to the production of a particular type of symptom with a complex array of both proximate and remote causes and mechanisms placed "upstream", both in the organism showing the disturbance and outside (for further distinctions see later).

The operation by which treatment effects used for *ex juvantibus* inferences are declared to be beneficial is largely ideological in many cases. Such treatments are seldom curative, or even palliative in the beneficial sense of the word, but often simply tools for symptom modification, for imposing some kind of control and for strengthening the position of the controllers. For example, several hundred million patients have been treated with neuroleptics since their introduction, often with high doses and for extended periods of time. It must now, however, be realized that the cost-benefit balance is far from being as favourable as was originally claimed. A high proportion of the patients are turned into neurological wrecks due to irreversible damage to the nervous system inflicted by the treatments. Even more strikingly, it appears that pharmacological abatement of symptoms may reduce the chances of spontaneous cure, as measured by the increased frequency of relapses.[8] (Since the purpose here is not a discussion of psychiatric matters, I cannot deal with other aspects of the damage caused by the use of medical models of behavioural disturbances, or other objections to the systematic use of drug treatments.) Therefore, not only is theory strongly subordinated to what is arbitrarily considered to be useful practice; but the same ideological bias also leads both to the choice of practices and to the construction of models which are useful for the justification of such practices and for the internalization of control mechanisms.

These phenomena are quite evident in psychiatry, considering not only the demand for symptom control, but also (i) the ever-shaky diagnostic constructions which are in perennial need of some kind of support; (ii) the value of reductionist biological models to account for dramatic events which might otherwise lead one to question the prevailing structure; and, last but not least, (iii) the need for "objective" criteria to conceal the arbitrary bases on which subjects are declared to be insane or sane, still sick or already recovered; nonrecoverable or recoverable; nonfunctional (e.g. from the viewpoint of working capacity) or functional, etc.[9] To avoid any misunderstanding, one must hasten to add that reductionist (reifying) explanations can be drawn equally easily from areas other than biology, such as psychoanalysis and sociology.

Historical evidence on the arbitrariness of Ex Juvantibus models

The *a posteriori* or *ex juvantibus* criterion has a very long history both from the viewpoint of diagnosis confirmation and from that of model building. For example, the uterine model of female hysteria has long been supported by two main types of arguments. The first was based on the superficial analogies between states of sexual excitement culminating in orgasm and some of the symptoms of hysteria. The second was based on the fact that hysteric behaviour could be suppressed by strong physical and/or chemical stimulation of the external genitalia – not a surprising result indeed, if one considers the wide range of behaviours which are amenable to suppression by strong punishment.

In other words, *ex juvantibus* reasoning developed in the past was so crude that it could have been extended to any situation in which a given intervention (e.g. a stunning blow on the head) could stop a given symptom (e.g. a complaint about an ingrowing nail). The very fact that a uterine dominance model of female hysteria had a remarkable success, while a brain hyperactivity model of disturbed nail growth never got a real chance to get established, shows the ideological nature of the whole process.

After the positivistic method started to exert its influence, however, the *ex juvantibus* logic was gradually refined, thanks to the increasing pathological, pathophysiological and eventually also biochemical and molecular evidence on the mechanisms of action of various treatments. Attention, from this point on, deserves to be focused on an intermediate period between the rise of a particular reductionist methodology (positivism) and the first substantial scientific advances bearing either (i) on the aetiology of important diseases (e.g. Pasteur's work on microbial agents of infection), or (ii) on appropriate strategies to assess the mechanisms of treatment effects in relation to structural and functional aspects of the organism's make-up (e.g. the work of Claude Bernard on the neuromuscular junction and the mechanism of action of curare). In fact it seems important to answer what appears to be a chicken-and-egg question: which came first, the systematic use of the *ex juvantibus* criterion in the building of the new pathology which characterizes nineteenth-century medicine, or sound scientific evidence on mechanisms of action, allowing the *ex juvantibus* logic to be placed on somewhat firmer ground? If the former preceded, rather than followed, the latter, then there is further evidence that the whole process was tainted with ideological elements, i.e. based on arbitrary choices serving all-too-easily identifiable purposes.

Since I am not a historian of the biomedical sciences, my own search in the area outlined above has necessarily been limited. Some time ago, however, I came across the Italian edition (1842) of the well-known *Traité de thérapeutique et de matière médicale* by A. Trousseau and H. Pidoux, first published in Paris in 1836–9. (Undoubtedly this treatise was influential, as shown by the ten successive editions, the last of which is dated 1877, and by the fact that it was translated in other languages). The book, otherwise of a very high quality relative to the time when it was published, contains many examples of an illegitimate use of *ex juvantibus* logic. It thus testifies that the fashion of drawing inferences about disease models on the basis of assumed mechanisms of action of treatments was quite widespread before any critical scientific evidence of the kind mentioned above had become available.

At least three main variations of this kind of logic can be identified in Trousseau and Pidoux. The first one uses as a starting point the effects of a treatment whose unquestionable effectiveness had been known for a long time, and was bound to be confirmed at a later time by the newer scientific methods. In the case of malarial fever treated by quinine, for example, knowledge about the aetiopathogenesis of the ailment was limited to empirical epidemiological evidence on the unhealthiness of marshes (and sometimes also on the dangerousness of insect vectors), while scientific evidence on the mechanism of action of quinine was practically nil. Trousseau and Pidoux exploit this void to assume a particular mechanism of action of the drug – restoration of the energetic stability of the organism – which allows them to promote a disease model assuming a periodic weakening of the vital resistance by the effluvia of the marshes. This, in turn, provides an appropriate basis for a bioenergetic model of normal functioning, which is in agreement with the substitution of the traditional (mechanistic) reductionist models with the newer (bioenergetic) models. Human beings, in other words, are more than Descartes's automata: a better analogy is that with the all-important source of industrial power, the steam engine.

A second and third version of *ex juvantibus* logic were based, respectively, on treatments which were subsequently shown to be devoid of any intrinsic efficacy (positive evaluations were probably based on placebo-type improvements and/or adventitious reinforcement phenomena), and on treatments which were certainly capable of modifying the course of diseases, but basically for the worse rather than for the better.

I have already mentioned a modern version of an *ex juvantibus* model based on treatments which, on average, can do more harm than good, that is, administration of neuroleptics to psychotics. Therefore it is interesting to take a look at the early nineteenth-century counterparts of such distorted uses of *ex juvantibus* logic. Quite understandably, the equivalents can be found in the realm of devastating infectious diseases with a poor prognosis, i.e. in situations in which the physician obeyed a *do-something* imperative in an attempt to cope with an ominous clinical picture.

One case handled by Trousseau and Pidoux is particularly striking. In fact, the authors insist that the physician must not give up emetic and cathartic treatments in certain kinds of ailments with fever and gastrointestinal symptoms, in spite of pathological evidence showing severe bowel lesions, particularly in the lymphoid structures of the intestinal walls. (This strongly suggests that they are speaking of typhoid). The benefits of what we now see as dreadful "therapies" were unequivocally shown in their opinion by the massive elimination of gastrointestinal contents which undoubtedly must have preceded the patients' breathing their last, while these "benefits", in turn, provided firm evidence for the "bilious" or "putrid" aetiology of the disease.

Such acrobatics, although strongly questioned by the rising schools of morbid anatomy and pathology, appear to rest both on the high death rate which was expected anyway, with or without treatment (hence the eagerness for some kind of intervention, *ut aliquid fieri videatur*), and on insufficient diagnostic criteria, allowing severe infections like typhoid to be confounded with benign, although dramatic, instances of gastrointestinal upset. In the latter type of cases, adult patients in good physical condition had a high chance of cure anyway; therefore, in a situation which is roughly amenable to a two-way classification (lethal versus benign ailment; treated versus not treated), it was probably easy to focus the attention on the apparently good results of treatments in patients whose prognosis was favourable anyway. The goal was confirmation at any cost of the power of the physician. The *ex juvantibus* logic was often pushed well beyond the level of a convenient disease model, to reach the level of an even more convenient model of normal functioning. When speaking of fever in general, for example, Trousseau and Pidoux make this method quite explicit: an adequate model of fever must rest on an adequate model of normal heat production, and vice versa. In other words, although the authors are sophisticated enough to avoid patching all illegitimate inferences into a single piece of reasoning, the general strategy is quite obvious. Assumed mechanisms of action of treatments which are useful (or are declared to be useful) must provide a good start in the direction of a convenient disease model. Subsequently, circular reasoning on models of disease and models of normal functioning must act both as a powerful fly-wheel in the logical process, and as further justification of the treatments themselves. Reification of people, both sick and healthy, is easily achieved by these pathways.

Another important result of such approaches was the possibility of attacking older doctrines which denied the specificity of lesions and disease processes, and which therefore did not lend themselves to the piecemeal approach to the problems of diseased subjects which is one of the main tools of expropriation and reification. (This, obviously, is not to support the view that the older models were correct, but to show that there was considerable haste in overextending the philoso-

phy associated with the new ones, to the point of denying any importance to the subject as such). Modern counterparts of this phenomenon are quite evident, for example, in the harsh attacks of biological
psychiatrists against field theories of mental illness which deny a
sharp qualitative separation of various disease entities. Further discussion of this important point, however, is not possible here.

Specificity of proposed solutions: time and place factors

These comparisons emphasize the fact that in different periods and/or
places one must cope with different contradictions, e.g. infectious
pathology yesterday versus non-infectious pathology, including widespread psychiatric disturbances, today.

Even more important for comparative purposes is the fact that a
given type of disturbance can be met with quite different buffering
systems, depending on the context. So, for example, the traditional
(appalling) versions of the asylum sufficed to cope with psychiatric
problems as long as they were not questioned by those who managed
to stay out of the asylum itself, and closed their eyes to what happened
to relatives and acquaintances who had been segregated. Such acceptance in turn was strictly related to the living conditions and harsh
rules by which people had to abide – for the great majority, hard man-
ual work, low wages, job insecurity, Spartan alimentary habits, mis-
erable housing, low-quality alcoholic beverages as the main if not the
only source of leisure, high rates of infant mortality, tuberculosis,
syphilis and other infectious diseases, and last but not least, army
charges and volleys of bullets at the slightest disturbance in the
streets. Segregation, drastic treatments, and social-Darwinistic models of insanity and criminality fitted well in this picture. With the
changes in living conditions brought about mainly by the struggles of
organized workers' movements, a greater sophistication became
necessary – revolving-door policies for mental hospitals, community
mental health services, biological treatments such as drugs and ECT
supported by a confounding amount of scientific data, and biomedical
models of disease based on more refined versions of the *ex juvantibus*
logic. The basic quality of the whole thing, however, has changed very
little, particularly since the long circular trip through environmentalist psychological models and various versions of dynamic and social
psychiatry has eventually resulted in landing at a basically biomedical
harbour, although with a variable amount of non-medical parapher-
nalia.[10]

Even in our times, reductionism is often fought nominally at the surface, but basically accepted in the substance. This is shown by the fact that few efforts are made to understand the ways by which deeply rooted logical systems were accepted, not to speak of the fact that adequate scientific and philosophical tools to replace the neopositivistic approach are still far from being available. Finally, it often happens that those who are more strongly against the reductionist logic cannot do away with the tools developed according to the logic itself; this applies, for example, to the use of psychopharmacological agents by politically active psychiatrists committed to the fight against the asylum. The consequence of this contradiction is often that a number of people among psychiatrists, nurses, social workers, patients, and their relatives and acquaintances, end up maintaining a purely medical view of mental ailments. (In fact the *ex juvantibus* logic is so deeply rooted that if a drug is the more visible tool among those employed to help the patients, then there must be a basic similarity between mental and other diseases. This in turn assigns to the patient a sick role which may well favour relapses and chronicity.)

Caveats, and typology of ideological biases

Before concluding, it is necessary to deal with some caveats in this analysis. In fact, when evaluating the *ex juvantibus* logic one must distinguish between the cautious empirical use in practice of an imprecise but still helpful instrument – e.g. for diagnostic purposes – when it is not feasible, or desirable, to mobilize the resources needed for research problems, and the use of the same instrument as the principal discriminant in establishing an explanatory model.

Furthermore, the use of the *ex juvantibus* criterion in the evaluation of the relative merits of alternative models is not illegitimate *per se*, but becomes illegitimate if one refuses to consider its severe limitations. More specifically, the *ex juvantibus* criterion not only does not allow any inference about remote causes, but also, in the absence of additional independent evidence, it cannot discriminate between at least four different types of inferences concerning (i) proximate aetiological factors, (ii) fundamental ("upstream") pathogenetic mechanisms, (iii) peripheral ("downstream") mechanisms responsible for the production of particular symptoms, and (iv) still other mechanisms which have little to do with those so far listed (an example: a problem child who has broken his leg and has been put in a plaster stops kicking people and furniture; this kind of symptom abatement does not allow us to draw any inference on underlying aetiopathogenetic factors and mechanisms of symptom production).

The distinction drawn above is far from having simply academic interest. For example, a genuine curative action of neuroleptics is now denied almost universally, with the result that one cannot reason *ex juvantibus* on aetiological factors in psychosis. Once it is acknowledged that the effect is no more than symptomatic, one should discriminate between two possible inferences based on symptom abatement and drug mechanism of action (e.g. antidopaminergic properties). The first is that symptoms, whatever the original cause of the ailment, are produced at least partly via a disturbance of dopaminergic mechanisms. The second is that dopaminergic mechanisms are involved in some kind of behaviour activation system which is responsible for the modulation of a wide range of behavioural outputs, both normal and abnormal.

As a matter of fact, a growing literature on the comparable efficacy of neuroleptics in ironing out both "desirable" and "undesirable" behaviours of psychotics, taken together with the evidence on unfavourable consequences of their administration, possibly resulting in a higher relapse rate and a higher risk of chronicity,[11] seem to support the second, i.e. the less favourable inference.

Furthermore, in spite of the growing awareness of these limitations, therapeutic decisions continue to be mainly in the direction of massive neuroleptic impregnation of innumerable patients for extended periods of time. In other words, theoretical choices and operational decisions made on an ideological basis can withstand evidence on their arbitrariness until more convenient choices and decisions can be made on the basis of a different, and equally arbitrary, ideology, or until mystification is no longer necessary.

The history of syphilis over the past hundred years provides a striking example of this process. Treatments which were practically ineffective and highly toxic, such as those employing mercury preparations, gave way several decades ago to the organic chemicals introduced by Ehrlich on the basis of his concept of *therapia magna sterilisans*. Undoubtedly, arsphenamine (Salvarsan) and neoarsphenamine (Neosalvarsan) were aetiological agents. However, it would not have been difficult to show (as it has been shown recently) that, *on average*, they did more harm than good, particularly in a context where the tendency was to ascribe all human troubles to syphilis, and diagnostic tools such as the serological test of Wassermann, with a very high rate of false positives, were used indiscriminately. So the decision to make a massive use of the compounds in a very large number of cases which were placed under strict medical control (and often resulted in severe intoxication and even death) was mainly ideological. This type of mystification ceased to be necessary when penicillin became available, since this drug, although not without dangers, combined highly specific aetiological properties with a highly favourable risk-benefit ratio. In other words, a more complete analysis could argue about the factors which made penicillin available at a particular time, as one of the results of the war effort, and about the factors which maintain a fairly high rate of venereal disease, thereby making it necessary to treat affected people with antibiotics. However, the treatment of each individual patient with the more appropriate antibiotic preparation cannot any more be defined as an arbitrary (ideological) choice, as was the case previously, e.g. with the use of mercury derivatives and organic arsenicals.

Returning for a moment to the neuroleptic example, one last remark is in order. Let us assume for a moment that the drugs could be proven to possess specificity either at the aetiopathogenetic level (proximate causes or fundamental disease mechanisms) or on symptom production (selective abatement of undesirable behaviours with no effects on desirable behaviours, to use a questionable but fashionable terminology). At this point it would still remain an arbitrary ideological choice to confound a *biological component* in a complex phenomenon with important historical, cultural, socio-economic and political *remote* causes (which is shown e.g. by the changes in the prevailing behavioural pathologies from place to place and in different times) and a *biological explanation* of the phenomenon itself, not necessarily at the level of nominally accepted explanations, but at least at the level of accepted methods of modification and control. (As emphasized before, if one subscribes to a nonbiological model of mental illness and then proceeds to employ drugs or even ECT as main tools in their treatment, the result will eventually be the same as the one obtained via biopsychiatric models). This mystification is quite frequent in medicine, as shown by the high rate of infectious pathology in Italy resulting in a very large number of antibiotic treatments. Strictly speaking, it is true (non-ideological) that microbial agents are proximate causes and that infected people are best treated by the appropriate antibiotic. The choice to leave remote causes where they are, however, has a precise meaning which needs no explanation here. This example, of course, constitutes an oversimplification relative to problems such as those which are encountered in psychiatry. However, it helps in understanding the close relations between the convenience of dealing with a given problem in a particular way and the decision to favour explanatory models (either in principle, or *de facto*) which, if not grossly incorrect, at least inflate one aspect until it occupies the whole perspective.[12]

The present discussion of the ideological choices behind *ex juventibus* reasoning and the operational decisions related to it can only scratch the surface of a large problem which has so far received insufficient attention. In Italy, in particular, the scientific left has experienced considerable difficulties in the discussion and demystification of bioreductionist approaches to human problems. This applies to the earlier phases of such a discussion in the second part of the nineteenth century, when many marxists took up whatever they could find in the newer scientific developments to fight obscurantism and reaction.[13] There were, of course, exceptions confirming the rule, as in the case of Antonio Labriola. Writing to Engels in 1891, he used a caricature of the social-Darwinistic jargon spread by Cesare Lombroso to affirm: "In my opinion, positivists are the representatives of the cretin degeneration of the bourgeois type".

Similar problems have been encountered in more recent years after the long pause due to the two world wars, fascism and the difficult post-war period. In fact, the eagerness to displace the fascist heritage as well as the spiritualistic and idealistic trends still controlling Italian culture and academe and opposing themselves to an adequate status of the sciences, has made neopositivistic scientism (or, at best, outdated Bernalian attitudes) dominate marxist thought in the past two decades. In the case of the behavioural sciences, for example, there has been little between total ignorance of the biological approaches and naive flirtation with ethological and psychobiological models applied to human problems.[14] In the few instances in which efforts were made to overcome this situation, difficulties were soon raised from several sides, and progress came to a halt.[15] Clear examples of the insufficiency of the debate can be found in recent discussions on sociobiology, which limit themselves to superficial criticism of its possible misuses, without any in-depth analysis of the basic flaws in the method and the logic.[16]

There are, however, growing signs of uneasiness about the dangers of such a situation, which leaves ample room both for the absorption of "scientific models" (i.e. ideologies) diffused by powerful sectors of the international scientific establishment, and for the justification of practices such as those aiming at behaviour control, which in turn provide further confirmation of the underlying ideologies. (See above the example of organic treatments administered by opponents of biological psychiatry, thus leading to internalization of the biomedical model all the same).

Such uneasiness, however, comes at a time of economic and political difficulties which have severe repercussions on the cultural, scientific and academic worlds marked by an increasing pressure to conform, often under nominal left-wing labels. As emphasized by Barker through the quotation from Hannah Arendt,[17] it is easy, under such circumstances, to promote a convenient image of humanity until people accept the transformation of themselves into this image, namely, to create for reductionist models a wide basis of consent. As a matter of fact, there is no doubt that in a situation of helplessness they can work better than any hard drug, sophisticated tranquillizer, or expert psychotherapy. This, obviously, does not apply only to biological reductionism, since modern science offers a wide range of reifying models among which individuals can choose to abate their anxiety and forget the history of their kind. As Bernard de Fréminville has appropriately summarized, "the manipulation and control of individuals requires techniques which have to be the more personalized, the greater the masses to be manipulated and controlled . . . It is easier to dispense blindfolds than to try to hide the sun."[18]

Notes and references

1. This paper is a modified version of an article I have recently published in *Sapere*, No. 818, pp. 2-14, April-May 1979, entitled "Pratica e ideologia del farmaco". Although the responsibility for the contents of the present article is entirely mine, I am especially indebted to the following people with whom I have worked and discussed for several years: Marina Frontali, Luciano Terrenato and Valerio Giardini, who are among the co-authors of the volume *Psicobiologia e Potere: Il nuovo Socialdarwinismo*, Milan 1977; Franca Ongaro Basaglia, who is the author of a series of articles in *Enciclopedia Einaudi* (Turin) which are relevant to the present discussion, since they deal among other things with ideological aspects of the history of medicine (*Clinica*, vol. 3, pp. 222-42, 1978; *Cura/Normalizzazione*, vol. 4, pp. 306-23, 1978; *Farmaco/Droga*, vol. 6, pp. 38-52, 1979; *Medicina/Medicalizzazione* – which I have co-authored – vol. 8, pp. 999-1041, 1979; Michele Risso, with whom I have co-edited and partly co-authored several materials on the mystification of technical intervention (including psychopharmacological and other organic treatments) in psychiatry (see *Fogli di Informazione*, special issue No. 57-58 on "Psicofarmaci", 1979).

2. M. Barker, "Biology and ideology: the uses of reductionism", in this volume.

3. This approach can be traced to ancient times, as shown in the analysis by Mario Vegetti, *Il Coltello e lo Stilo: Animali, Schiavi, Barbari, Donne, alle Origini della Razionalità Scientifica,* Milan 1979.

4. See the work by Franca Ongaro Basaglia quoted in note 1.

5. P. Achard *et al., Discours Biologique et Ordre Social,* Paris 1977.

6. Although couched in a more cautious language, several well-known analyses published in English testify on this point, coming from authorities such as R. Dubos (*Man Adapting,* 1965) and T. McKeown (*The Role of Medicine: Dream, Mirage or Nemesis,* Nuffield Provincial Hospital Trust, 1976). See also "Medicina/ Medicalizzazione" in *Enciclopedia Einaudi,* quoted in note 1.

7. See, in the volume *Psicobiologia e Potere* quoted in note 1, G. Bignami and V. Giardini, "Le 'basi scientifiche' della psichiatria biologica," pp. 95-127.

8. L. R. Mosher, *Soteria's Model of Madness: Implication for Treatment,* Rockville 1977; M. Rapoport *et al.,* "Are there schizophrenics for whom drugs may be unnecessary or contraindicated?", in *International Pharmacopsychiatry,* 1978, vol. 13, pp. 100-111; J. W. Perry, "La psicosi come stato visionario", in *Rivista di Psicologia Analitica,* vol. 17, pp. 213-21; S. M. Matthews *et al.,* "A non-neuroleptic treatment for schizophrenia Analysis of the two-year post discharge risk of relapse", in *Schizophrenia Bulletin.* Furthermore, schizophrenic patients have a much lower relapse and chronicity rate in developing countries (where treatment and the corresponding ideologies are to a considerable extent non-medical) than in industrialized countries (see W. H. O. *Schizophrenia – An International Pilot Study,* Chichester 1979).

9. These comments on the arbitrariness of nosographic, diagnostic, and functionality criteria obviously reflecting ideological, socio-economic and political needs, should not be equated with statements such as "psychiatric illness does not exist", etc. Considering, however, the evidence on historical changes of psychiatric phenomenology, as well as the impressive data on the consequences of labelling, on chronicity induced by hospitalization (and possibly also by indiscriminate symptom abatement), one is tempted to conclude that the greatest portion of the overall variance should be ascribed to models and practices with an arbitrary ideological basis, while a much smaller portion appears to be due to mental illness *per se.*

10. In fact, reifying models of a non-biological kind (e.g. those developed by psychoanalysis and sociology) have been widely used for the internalization of control mechanisms and for the diffusion of convenient explanations of psychological suffering. In the long run, however, the models and tools for control provided by the non-medical sectors, although useful as auxiliaries, have ended up being non-competitive with those provided by the biomedical sectors. As indicated by the policy decisions of the responsible agencies in the US in recent years, or the campaign in Great Britain to rehabilitate ECT in the eyes not only of specialists (who never hesitated to use it), but also a wider public, the "hard core" becomes more and more a medical one. Supplementary tools – networks of social workers, behavioural engineering techniques based on either operant or classical conditioning models, supportive psychotherapy for the many and more intense (psychoanalytical, or other personal support for the few) – have to be provided by several of the non-medical disciplines which maintain important areas of influence, but are constrained in a subordinate role.

11. See note 8.

12. This can be seen also at down-to-earth levels, as when the authorities made heroic efforts to blame mussels (and Neapolitans eating them raw) as the main problem in the cholera outbreak. Also, to reduce the emphasis on the dramatic socioeconomic conditions in Naples during the recent outbreak of fatal respiratory ailments in small children, virological investigations by local, national and international experts were turned into a thriller, while useless though spectacular interventions, including those by military forces, were staged *ut aliquid fieri videatur*.

13. An impressive documentation on this point can be found in the book by L. Bulferetti, *Le Ideologie Socialistiche in Italia nell'Eta del Positivismo Evoluzionistico (1870–1892)*, Florence 1951. A concise analysis of the confusion which reigned in the logic of Cesare Lombroso, allowing him to be a fervent socialist while developing his social-Darwinistic models and the corresponding tools for behaviour control, can be found in the introduction (by F. Giacanelli) to the book by G. Colombo, *La Scienza Infelice – Il Museo di Antropologia Museo di Antropologia Criminale di Cesare Lombroso*, Turin 1975. More cautious evaluations on these points can be found in the well-known monograph on Lombroso by L. Bulferetti, written several years after the more aggressive study quoted above. Quite significantly, a Lombroso revival is in progress at the international level, as shown by recent meetings and initiatives for the reappraisal of his work.

14. I have written on this point several articles in the daily *Il Manifesto*, which would be out of place in a regular bibliography. I hope, however, that within a year or so from this meeting I will be

able to write a more systematic article or book chapter on this delicate topic.

15. This, for example, happened when part of the reports and much of the discussion at a meeting on "L'uomo d'oggi fra natura e storia" (humanity today between nature and history), held at the Istituto Gramsci 1978, January 27-29, went somewhat beyond the boundaries which are usually respected in the criticism of science by the so-called official left in Italy. In fact, the two introductory reports by G. Berlinguer and M. Aloisi appeared at about the same time in *Critica Marxista*, and the first of these was soon republished as part of a book. However, the engagement to publish a volume of proceedings c/o Editori Riuniti (the publishing house of the Italian Communist Party) has not been met to the date of this writing (1 March 1980), nor have the participants been given any plausible explanation for the delay.

16. See the debates published in *L'Unità* (24 January 1980, p. 8) and Rinascità (15 February 1980, pp. 20-22), and the introduction to the Italian translation of Wilson's *Sociobiology* (Bologna 1979), to be confronted with the criticisms available in English, e.g. by members of the Boston sociobiology group of Science for the People. (Detailed quotation of the materials discussed by other participants at the meeting is omitted here.)

17. See M. Barker's chapter in this volume.

18. B. De Fréminville, "I laboratori dell'interiorizzazione", in *Fogli di Informazione*, no. 57/58, pp. 332-6 (for further information on this journal, see note 1).

7
Hierarchical Structures and Structural Descriptions
Giacomo Gava

The picture which today science offers us of itself is a complex one. Its fields of research are almost unlimited and they incessantly increase and proliferate in such a way that it would be very difficult and, to a certain extent, useless for the topic of this paper to present a complete and faithful image. Think of the host of scientific departments, research groups, specialized institutes and laboratories spread all over the world, representing all the branches of science with their laws and theories which are supposed to describe and explain the different domains of reality. Although any simplified and linearly ordered model is inadequate to understand this whole complexity, it is perhaps useful to start from an oversimplified model illustrating the most fundamental structural levels and their corresponding structural descriptions and explanations and then to show their internal, tangled branching.

From a structural point of view living and non-living beings in the world can be considered on the basis of the following different levels: the macroscopic or molar level, the neurophysiological, anatomical, biochemical and chemical level, and the microscopic and sub-microscopic level. The first level is characterized by observable behaviour, functions and so forth. The second level comprehends all neurosciences, from the chemical and biochemical to the neurophysiological and physiological ones. The microscopic level is concerned with the quantum domain of atoms and molecules and the sub-microscopic one deals with the so-called elementary particles. To the above structural, hierarchically-ordered levels different language strata correspond: ordinary language; religious, philosophical, political, economic, social, psychological etc. languages; neuroscience languages; and physics languages. Each of these languages has its own laws or rules and each refers to a certain domain of objects. In "Unity of Science as a Working Hypothesis" Oppenheim and Putnam[1] advanced the following scheme of historical structural levels to which six "fundamental disciplines" correspond, social groups, (multicellular) living things, cells, molecules, atoms, elementary particles. They recognized the oversimplification of their proposed model and that the structure of science is more complicated. Of course, several other hierarchies of organization or models could be mentioned. A new sort of complexity, however, arises when we come to deal with the various aspects and/or levels of phenomena and/or problems. Aphasias, for instance, are the objects of study not only for neurologists, but also for psychologists, physicians, linguists, speech pathologists, educators, anthropologists, philosophers and many others. As can be imagined, the hierarchical structures and the corresponding levels of language are potentially innumerable. Yet however numerous these distinctions may be, they will always remain finite, relative and not absolute, for matter is limited and we are not gods.

Besides, within the whole hierarchical system the relationships do not take place only, so to speak, horizontally but also extensively, vertically and/or diagonally and dynamically. In other words, we can discern in it four fundamental dimensions which need exploration: the first is given by the correspondence between the single structural levels and their respective levels of language; the second is constituted on the one hand by the interactions among the components of the same structural level, and on the other hand by the connections among the different laws or rules and theories of a particular language; the third is the knotty problem of the interrelations between different structural and language levels; and the last springs from the evolution and dynamics of languages, their ever-increasing terms and the constant changing of their meanings, and the underlying dynamical and evolutionary processes which they describe. To all this we can add the method of investigation, which can be ontological, methodological, epistemological or logical, approaches which are not necessarily taken separately.

The hierarchical levels, whichever order is proposed, are conventional, largely due to the impossibility of our human brain grasping the real objects both in their totality and in their details. Frequently in the past humans have tried to explain reality in an unitary way – think of the various forms of pantheism and of vitalism – but this was some metaphysical Weltanschauung, completely empty of informative content and of cognitive, factual meaning, and unable to predict observable events and behaviour. The limits of the brain are most probably determined by, among other factors, its present stage of evolution, by the firing of its neurons and by their functions not being fully exploited by us. Whatever the cause, at the present moment we seem unable to cross a certain threshold in understanding in a unified way both the details and the abstract and general traits of things. And this procedure is now the most practical to deal with the things of nature. Only very few people are specialized in more than one field of science.

So, as a result of the human brain's long evolution, we find an indefinite number of languages coexisting, mutually and continuously interacting in various ways and making up the interwoven texture with which we are all acquainted. In particular, we observe that ordinary language— which is the direct and concrete expression of humanity in all our manifold and multifarious existential manifestations, which is conventional, which in many ways dominates human thinking through the continuous reinforcement of daily experience, and which still remains the most useful for practical purposes— constitutes in a more or less extensive manner the basic ground, the substratum of all other languages, which in their turn modify and develop with their overstepping terms such a language. Their permeating concepts and mutual and dynamical interactions make their boundaries not at all clear-cut, but undefined and partly overlapping. And since, as I have said, different dynamic logics and theoretical grammars govern the different languages, these too are partially overlapping. In short, every scientific discipline, which applies its own methods and logic, overlaps in a different quantitative way with other disciplines and their logics. And it is just these different logics that, according to many, make impossible complete translation from one language into another, for instance, from ordinary language into formal language (Ryle, "Formal and Informal Logic"[2]) or from one scientific language into another.

Furthermore, every language has its own history, which very often is not easy to reconstruct adequately. Let's give a few summary examples. Not taking into account ordinary language, which is the oldest, philosophical language, as has been pointed out also by ordinary language philosophy, arose out of ordinary language. But we cannot fully accept any longer this "ordinary language" enterprise, since many relevant philosophical problems find their genesis in scientific developments and theories— think, for instance, of the epistemologies worked out by Popper, Kuhn, Lakatos and Feyerabend and of their controversies. Nevertheless, ordinary language remains the primary component of philosophical language, which in its turn is constituted by different logics, according to its internal systems. Psychological language has its origin in the sixteenth and seventeenth centuries and became sufficiently autonomous in the second half of the nineteenth century. At first it was closely connected with philosophy, but in the long run its dependence grew weaker. Now it incorporates to a certain degree the inherited philosophical language and, partly due to this, the ordinary language underlying it, besides the methods coming from other sciences. Neurophysiological language, which is tightly linked to all other neurosciences, is quite recent – compare it, for example, with the long-established ordinary or philosophical language. We can say that it started off in the second half of the last century and, notwithstanding its significant progress, its incompleteness (which should not surprise anyone) is chiefly due to its new appearance.

Since among scientists and philosophers there are many divergent opinions and controversies on the dimensions I have discussed which seem to come down to the major problem of reduction or non-reduction, I think it useful to recall here briefly some of the main ones. Carnap in two famous works (1936-7, 1938), [3,4] improving on his preceding and more radical formulation of the thesis of physicalism, doesn't speak any longer of complete verification, of translatability of the terms of all branches of science into a physicalistic language, but only of "a process of gradually increasing confirmation" (ref. 3, p. 134) and of reducibility, testability and confirmability of such terms. His more liberal epistemological requirement of empiricism is put like this: " 'Every synthetic sentence must be confirmable'. . . Predicates which are confirmable but not testable are admitted; and generalized sentences are admitted" (ref. 3, p. 189). Interested as he is in the logic of science and not in ontology, in the relations among the laws of the various branches of science, he argues that in science there is "*no unity of laws*", yet: "It is obvious that, at the present time, laws of psychology and social science cannot be derived from those of biology and physics. On the other hand, no scientific reason is known for the assumption that such a derivation should be in principle and forever impossible" (ref. 4, p. 422). However, there is a unity of method, of language, of the "physical thing-language", whose unity is "a necessary preliminary condition for the unity of laws" (ibid.).

A unification programme of science based on successive microreductions is defended by Oppenheim and Putnam.[1] They distinguish three concepts of unity of science: unity of language, unity of laws, and unity of laws within the system of the reducing branch of science. Unity of laws "is attained to the extent to which the laws of science become reduced to the laws of some one discipline" (ref. 1, p. 4). But they add: "Unity of science in the strongest sense is realized if the laws of science are not only reduced to the laws of some one discipline, but the laws of that discipline are in some intuitive sense 'unified' or 'connected'. It is difficult to see how this last requirement can be made precise; and it will not be imposed here" (ibid.). So they claim that the proper current method of unification lies in microreduction, according to which "the branch B_1 deals with the parts of the objects dealt with by B_2 . . . and the objects in the universe of discourse of B_2 are wholes which possess a decomposition . . . into proper parts all of which belong to the universe of discourse of B_1" (ref. 1, p. 6), and hence the successive microreduction of all levels to the "unique lowest level" which is constituted by elementary particles. At present, there is no unitary science but they consider it a working hypothesis "credible . . . on methodological grounds" and on "a large mass of direct and indirect evidence in its favor" (ref. 1, p. 28), of which their article furnishes some salient examples.

Nagel, [5,6] following the empiricist tradition, works out two necessary and sufficient conditions or criteria for translating one science into another: the "condition of connectability" and the "condition of derivability". The first states: "Assumptions of some kind must be introduced which postulate suitable relations between whatever is signified by 'A' and traits represented by theoretical terms already present in the primary science" (ref. 6, p. 353-4). "A" is a term which refers to something absent from the primary science. This condition requires that the "*theoretical* terms of the primary science appear in the statement of these additional assumptions" (ref. 6, p. 354, note 4), it is in general not considered sufficient for reduction, and so it must be "supplemented" by the second condition, which is formulated in this way. "All the laws of the secondary science, including those containing the term 'A', must be logically derivable from the theoretical premises and their associated coordinating definitions in the primary discipline" (ref. 6, p. 354). A few pages later Nagel explains also that such a reduction neither eliminates nor reduces to something unessential the distinctions and the types of behaviour recognized by the secondary discipline.

Strong objections to Nagel's theory of reduction were raised by Feyerabend. [7,8,9] He uses "the *consistency condition*" among successive theories and "the *condition of meaning invariance*", two criteria which Nagel should accept. Feyerabend lingers particularly on the change of meaning of the terms belonging to theories involved in the process of reduction. The concept of "mass", he argues, according to whether it is used in classical mechanics or in the theory of relativity, has two different meanings. Consequently, he rejects their reduction; and his solution consists in eliminating and replacing the terms once we discover that they designate entities which do not exist. Here reduction becomes even more difficult to realize on account of his defence of an ever increasing "theoretical pluralism" of alternative, incompatible, incommensurable and competing theories in science. Then he writes: "the invention of alternatives in addition to the view that stands in the centre of discussion constitutes an essential part of the empirical method", and seeks a little further on "method that is compatible with a humanistic outlook" (ref. 8, pp. 176 and 179). (Some criticisms and developments of these positions of Nagel and Feyerabend together with a short bibliography on these matters can be found in Lévy[10]).

Simon[11] attempts to reconcile what he calls the "Laplacian-Mendelian approach", that is, reductionism with non-reductionism. For him, each dynamic complex system of nature is made up of interacting subparts; it exhibits an underlying hierarchical structure and it is described as a "nearly decomposable system". The components of every higher level have only a certain degree of independence of their microscopic components. The properties of the lower levels do not appear completely on the upper levels: their exactness and precision are sacrificed. So a higher level has explanatory power lacking on the lower level and reduction of one level of explanation to a more basic one does not render the first explanation "otiose or dispensable" (ref. 11, pp. 179, 258-61 and 270). "Scientific knowledge is organized in levels, not because reduction in principle is impossible, but because nature is organized in levels, and the pattern at each level is most clearly discerned by abstracting from the detail of the levels far below . . . And nature is organized in levels because hierarchic structures – systems of Chinese boxes – provide the most viable form for any system of even moderate complexity" (ref. 11, pp. 260-1). This trend of thought was recently followed by Hofstadter (ref. 12, pp. 520-22, 706-19), who, however, "enriches" it with the emergent theory elaborated by Sperry.[13]

A similar view was also held by Pattee [14,15,16] with reference to hierarchical organization in biology, from molecule to brain. He agrees neither with Crick[17] nor with Polanyi,[18] who are taken as two extreme examples of reductionism and non-reductionism respectively in biology. The hierarchical constraints, which are the "alternative descriptions" of a system, are "non-holonomic" or "non-integrable". The structural hierarchies are given by a "loss of degrees of freedom" at the most detailed level. For example, it is impossible to control all the molecules which make up a door, to control each degree of freedom; so we apply to an alternative description with fewer variables, to a more useful level of representation which, leaving out details, takes into account simplified and significant features. By virtue of this, the distinction between living and non-living matter depends on language-constraints and life cannot be reduced to "nothing but a very complex physical system" (Pattee, ref. 16, p. 252), which would even go against most interpretations of modern physics. He affirms: "Life is distinguished from inanimate matter by exceptional dynamical constraints or controls which have no clear physical explanation" (ref. 15, p. 135).

Lévy, analysing the reduction of chemistry to physics, finds inadequate the "classical concept of reduction"[10] with its asymmetrical relation and whose origins she traces back to Nagel. She proposes the new concept of "reduction by synthesis". Such a reduction, which is a partial but adequate explanation of chemistry through physics, is attained by introducing "bridge-disciplines" or intermediary theories T_3, whose function is to establish connections between and to enrich the two disciplines; it is realized by "chemistry-physics models" which embody theoretical principles of physics and hypotheses of chemistry, and which are the meeting and exchanging point between the two sciences.

Lately, Rose [19,20] has tried to show how dialectical materialism "transcends" "reductionism as methodology and the ontological reductionism of identity theory". In analysing the mind-brain-cell problem, he accepts the classical distinction between structural levels and structural descriptions, and argues that the levels are operationally and "epistemologically distinct but ontologically unitary" – they may refer to the same objects. The hierarchy of levels is asymmetrical – upward and downward directions indicate respectively more complexity and more "basic" – and each level tends to be epistemologically complete. Among the different structural descriptions "there is not a causal, but a mapping relationship" – which is isomorphic – of "non-reductive identity". The translation between them (the languages examined are physiology and biochemistry) can be accomplished but in a "limited way"; in other words, not all statements of a language are translatable into the other.

The Identity Theory of Armstrong and others maintains that mental states or events are identical with brain or neural processes. The two sets of expressions do not have the same meaning and logic, but their logical independence doesn't entail their ontological independence, indeed they happen to have the same referents or denotata. It is a question of empirical identification: "Each individual mental process is in fact a brain process".[21] In this identity there is not implied any translation. Place does not sustain that "statements about sensations and mental images are reducible to or analysable into statements about brain processes" (ref. 22, p. 102); Smart asserts that "the thesis does not claim that sensation statements can be *translated* into statements about brain processes" (ref. 23, p. 163); and even Feigl, who adopts Frege's distinction between "sense" (which is different for both sets of expressions) and "reference" (which is the same for both of them), is of the same opinion: "But if this theory is understood as holding a *logical translatability* (analytic transformability) of statements in the one language into statements in the other, this will certainly not do" (ref. 24, p. 390). Rorty[25] defends the disappearance or replacement form of the identity theory, to which even Smart and Feigl afterwards adhered, at least partly.

There is no place here for detailed criticisms of the above positions, but we can observe that all of them fall roughly either into a reductionist side or into a non-reductionist side. To borrow an expression from Lakatos, we are here confronted with two main scientific research programmes, with several internal diversifications (for some further kinds of reductionism, in particular, and for some useful comments one can see Somenzi[26]). This state of affairs has led to more than one sceptical view about the possibility of filling the gulf existing between the various levels of language, to arrive at a unity of science. But in my opinion all this simply shows that the solution of these problems is anything but definitive.

The attempts at conciliation would seem to suggest that the solution has to be sought in some form of integration of the two systems, in a synthesis of the type realized *mutatis mutandis* by Newton and Maxwell, for instance: we remember how Newton's theory of gravitation unified both Galileo's terrestrial motions and Kepler's celestial motions, and how the differential equations of Maxwell accomplished the remarkable connection and explained the interdependence of magnetic field and electrical field. But I consider this enterprise not only unsuccessful but also impossible, because whereas the two latter syntheses bring together theories and laws within a single domain of science or at least within two closely related domains, in reduction or in non-reduction of all descriptive hierarchies and of the underlying dynamical processes two *opposite* epistemological conceptions are at stake. It is a question of logic for, strictly speaking, reductionism and non-reductionism do not allow a compromise: they are logically incompatible. I think it is more plausible to say that in the long run of future scientific development one program will overcome and supersede the other.

Looking at the history of science, however, we notice that dualistic approaches seem destined in due course to disappear – consider, for example, the many forms of vitalism and animism – and scientists in their laboratories apply a reductive methodology. There is a great deal of evidence that science strives towards a more wary and elaborated reductionism by successive steps, and that it is directed towards a reductionism in which the human brain will be at its ease in "level"-crossing and in which it will be able, always on the basis of empirically confirmable and predictable evidence, by means of newly developed theories, laws and instruments (computers), to catch the unity of things in the world and at the same time their internal, dynamical relations and interactive networks. To speed up the process, to reach this goal, a great help, I think, can come from the identity theory – especially if suitably matched with structural function analysis – which has already succeeded in making clear the relationships existing between ordinary language and above all neurophysiological language, and which has drawn attention to the possibility of replacing the first with the more explanatory power of the second, for all its present incompleteness.

In recent years we have noticed on the part of many advanced operators a revival of interactions among all the sciences – neurophysiology, psychology, linguistics, medicine, anthropology, evolutionary epistemology, history and philosophy of science, among others – which in some way or other are concerned with the brain and its different manifestations and problems. Such a mutual co-operation or interdisciplinary need is particularly felt today in the fields of physics, biophysics, anatomy, neuroanatomy, chemistry, biochemistry, neurochemistry, neurology, neurophysiology, physiology, biology, neurobiology and pharmacology (see, for instance, *Scientific American*, September 1979 and the recent meetings on the mind-body problem). Anyhow, we can "hazard" some predictions which make sense. It is reasonable, and not just a mere hope or a vague prediction, to state that deep-level explanations will survive and even supplant the

more misleading and ambiguous ones. In a preceding article (Gava,
more misleading and ambiguous ones. In a preceding article[27]
I pointed out that when psychology tries to fill the many gaps
by D. C. Dennett, then the above-mentioned kind of psychology is
doomed to die away, as neurophysiological advances will supply us
with exhaustive explanations of the human brain states, events and
processes.

In "suggesting" such a unitary goal, which is political and ideological only in so far as all science is political and ideological, my main
purpose has been to stimulate further inquiry and investigation in the
hastening of scientific development, in the rational conviction that, at
a more advanced stage, it will coincide with an unitary system transcending both non-reductionism and present reductionism.

References

1. P. Oppenheim and H. Putnam, "Unity of Science as a Working
   Hypothesis", in Feigel *et al.* (eds.), *Minnesota Studies in the
   Philosophy of Science*, vol. 2, Minneapolis 1958.

2. G. Ryle, *Dilemmas*, Cambridge 1954.

3. R. Carnap, "Testability and Meaning", in *Philosophy of Science*,
   no. 3, 1936, pp. 419-71 and no. 4, 1937, pp. 2-40; reprinted in R.
   R. Ammerman (ed.), *Classics of Analytic Philosophy*, New York
   1965.

4. R. Carnap, "Logical Foundations of the Unity of Science", in
   International Encyclopedia of Unified Science, vol. I, no. 1, 1938,
   pp. 42-62; reprinted in H. Feigl and W. Sellars (eds.), *Readings
   in Philosophical Analysis*, New York 1949.

5. E. Nagel, "The Meaning of Reduction in the Natural Sciences",
   reprinted in A. C. Danto and Sidney Morgenbesser (eds.),
   Philosophy of Science, New York 1960.

6. E. Nagel, *The Structure of Science*, London 1961.

7. P. K. Feyerabend, "On the 'Meaning' of Scientific Terms", in
   The Journal of Philosophy, 1965, pp. 266-74.

8. P. K. Feyerabend, "Problems of Empiricism", in R. G. Colodny
   (ed.), *Beyond the Edge of Certainty*, Englewood Cliffs 1965.

9. P. K. Feyerabend, *Against Method*, London 1975.

10. M. Lévy, "Les relations entre chimie et physique et le problème
    de la reduction", in *Epistemologia*, 1979, pp. 337-69.

11. H. A. Simon, "The Architecture of Complexity", in *Proc. Amer. Philos. Soc.* no. 106, 1962; reprinted in H. A. Simon, *The Sciences of the Artificial*, Cambridge, Mass., 1969.

12. D. R. Hofstadter, *Gödel, Escher, Bach: an Eternal Golden Braid*, Hassocks 1979.

13. R. Sperry, "Mind, Brain, and Humanist Values", in J. R. Platt (ed.), *New Views on the Nature of Man*, Chicago 1965.

14. H. H. Pattee, "The Physical Basis of Coding and Reliability in Biological Evolution", in C. H. Waddington (ed.), *Towards a Theoretical Biology*, vol. 1, Edinburgh 1968.

15. H. H. Pattee, "The Problem of Biological Hierarchy", in ibid., vol. 3, 1970.

16. H. H. Pattee, "Laws and Constraints, Symbols and Languages", in ibid., vol. 4, 1972.

17. F. Crick, *Of Molecules and Men*, Seattle 1966.

18. M. Polanyi, "Life's Irreducible Structure", in *Science*, no. 160, 1968, pp. 1308-12.

19. S. P. R. Rose, *The Conscious Brain*, Harmondsworth 1976.

20. S. P. R. Rose, "From causations to translations: a dialectical solution to a reductionist enigma", in *Towards a Liberatory Biology* (companion volume), London 1981

21. D. M. Armstrong, "Between Matter and Mind", in *Times Literary Supplement*, 17 February, 1978.

22. U. T. Place, "Is Consciousness a Brain Process?", in *British Journal of Psychology*, 1956, pp. 44-50; reprinted in V. C. Chappell (ed.), *The Philosophy of Mind*, Englewood Cliffs 1962.

23. J. J. C. Smart, "Sensations and Brain Processes", in *Philosophical Review*, 1959, pp. 141-156; reprinted in V. C. Chappell (ed.), *The Philosophy of Mind*, Engelwood Cliffs 1962.

24. H. Feigl, "The 'Mental' and the 'Physical'", in H. Feigl, M. Scriven and G. Maxwell (eds.), *Concepts, Theories, and the Mind-Body Problem*, Minneapolis 1958.

25. R. Rorty, "Mind-Body Identity, Privacy and Categories", in *Review of Metaphysics*, no. 19, 1965/6, pp.24-54.

26. V. Somenzi, "Philosophy, Biological Sciences and Cybernetics", intervention in *Congresso internazionale dell'associazione internazionale dei professori di filosofia*, Fondazione Cini, Venezia (in press 1979).

27. G. Gava, "Some Remarks on Psychology and Neurophysiology", in *Methodos* (in press, 1980).

8

Holism and Reductionism are Compatible

Allan Muir

Mechanism

I'll begin with some basic remarks concerning Newtonian mechanics, the paradigm for mechanism in general. This is employed as a concrete instance of ideas which have greater generality, for it already exhibits the limitations of mechanism in embryo form. These limitations become chronic in the analysis of complex systems, when the openness of the sub-systems becomes paramount.

Consider then the simplest mechanical system – a single particle moving on a line under the action of a constant force F. The motion is governed by Newton's equation $F = m\ddot{x}$, where m is the particle's mass and x its distance from a fixed point of the line; dots denote time derivatives. The system is deterministic in that the particle's position at any time t is fully determined by its position a and its velocity v at a given initial time t_0.

This simple instance exemplifies an important feature of physical laws, that they are conditional statements of the form "if a, then b". In our case, "if mass = m, force = F, position and velocity at time t_0 are a and v respectively, then position at time t is x". It is, of course, this conditional nature which affords their power and flexibility, but what we want to emphasize here is the corresponding requirement that the most mechanistic of systems is open. Closure is therefore a conceptual simplification or, less legitimately, a metaphysical assumption that some largest conceivable system – the universe – exists.

There are, of course, physical principles which, in our current state of knowledge, appear unconditional: for instance, the upper bound to the velocity of matter given by the speed of light and the limitations described by the uncertainty principle and the laws of thermodynamics. Such laws, however, are perhaps better seen as setting ontological boundaries – negative temperatures, supra-light speeds, exact locations in phase space do not exist – rather than as causal laws operating within the arena so delineated.

Our simple example serves also to illustrate the different characters of the various parameters. Apparently unchanging quantities intrinsic to a system, like m in the above, are accorded a differing status from the externally imposed force F.

Conceptual closure amounts to an assumption of constancy for the external factors, or at least a sufficient regularity to permit their phenomenological description. In such an analysis attention is focused on the internal mechanism and processes, with the exterior being represented by given forces, parameter values and so on. Enlarging the system to account for the source of these contingencies incorporates them into a wider mechanical model, but again there will be unexplained elements arising from the outside.

Consider, for example, a servo-mechanism to control a variable at a pre-set level – for concreteness one might imagine angular velocity controlled by a mechanical governor. The setting of the desired level is outside the system but a more encompassing model might reveal a mechanism for the setting which was itself controlled at a higher level. Indeed, from the standpoint of that higher level the details of the governor's action might well be irrelevant. It could be "black-boxed" and characterized merely by a sufficient input/output description.

Naturally these considerations are not confined to classical mechanics. Contemporary mechanism is well expressed through the concept of *dynamical system*. This is, in effect, a system of differential equations with a unique solution trajectory through each state point, permitting prediction from an initial state to the state at any subsequent time. Besides its classical mechanical uses this is a most popular scheme for mathematical modelling of biological systems with applications to nerve dynamics, morphogenesis, population interactions, enzyme kinetics and so on.[1]

The success of such modelling, at least in organizing concepts if not yet in prediction (using the word in the temporal sense, rather than in the sense of hypothesis generation), is most impressive evidence for a mechanical materialist philosophy. Indeed, I would argue that a major contribution of a dialectical alternative is to merely remind ourselves of the actual openness of all systems, which is expressed in mechanistic models through parameters and functions which phenomenologically describe the exterior.

Conceptual closure can fulfil two functions; either that of collapsing the exterior into a simple characterization, when attention is focused on the internal mechanism, or that of black-boxing the interior, when the intervention of the exterior is the principal concern.

The efficacy of such analytical strategies is the principal justification for conceiving the world as structured into levels. The reductionism/holism controversy becomes a non-issue whenever adequate conditions permitting closure of a system are formulated. For reductionism merely insists that higher level variables be expressible in terms of lower ones without the importation of emergent factors, while holism reminds us that each part of a system is constrained by the context of the whole.

Two kinds of hierarchy

Making these two views compatible in any given situation – the hierarchy problem – can be tackled in two essentially different ways.

Social hierarchies, for instance, are often described as a structure of commands, and tracing this structure is thought to be a sufficient explanation of any coherent behaviour. But the command hierarchy *par excellence* which is rapidly becoming an exemplar for all such hierarchies is the operating system in a computer. The scheduling of processor resources between concurrently running processes, under constraints dictated by priorities and costs, necessitates the creation of a hierarchy of structures linking the interface with a user to the interface with machine hardware.

The language employed for describing the required relationships of command within systems and programs is replete with terms like "priority", "slave", "control", "privilege". The nightmarish image which this evokes, of Daleks screaming orders to kill, might not be just a joke. Even if this were an appropriate language for the systems concerned, it would not be an entirely ridiculous enterprise to examine the relations between ideology and language in this context. A paradigm is being constructed which will inevitably invade other domains, such as neuropsychology and the organization of social life.

Turning to the other meaning of hierarchy, we find the word employed in natural sciences to characterize levels of material organization. The kind of language employed is quite distinct from that above. It would be rather bizarre to refer to a molecule as commanding its constituent atoms into a particular configuration. A physicist might prefer the word "constraining", but constraints are normally externally imposed. We can ask under what conditions a system might determine its own constraints, so that it holds together in a stable self-organization. Let's refer to such a structure as an autonomous hierarchy.

The gulf between the two approaches seems unbridgeable, yet to understand an object like the brain requires precisely this problem to be addressed. In its material aspect it shows a hierarchy of levels, through functional regions, neural masses, neurons, biochemical complexes and so on down into physical levels. At the same time the brain is an organizer of action within the external context. Whether this be analysed behaviourally, in terms of cognitive structures or however, we find a hierarchy of the command type.

In an intriguing discussion of the hierarchy problem Pattee[2] asks how an autonomous hierarchy could arise from a set of interacting components obeying physical laws. If the materials of which biological systems are composed are none other than those of inanimate physics, how can such a system behave in a coherent way? Coherence, as argued by Pattee, requires the totality of components to constrain each individual one, but the laws of physics are deterministic so surely they have no room for constraints, other than those externally imposed. Stated otherwise, the uniqueness of solution of the dynamical equations gives no leeway for other conditions to be applied. The state of motion just *is*; there is no sensible factoring of it into a constraint plus motion within the constraint.

Pattee goes on to argue that the idea of a constraint derives from the existence of an observer, who conceptually simplifies a complex set of interconnections into an easily stated condition. Thus a constraint is part of a description of the system, rather than a characteristic of the system itself. This leads him to say that if a system is to exhibit autonomous coherence, it must contain a description of itself; the system must be complex enough to register its own state.

The trouble with this argument is that it jumps too readily from autonomous to command hierarchy and so effectively dodges the issue of how coherence is achieved mechanistically. However, a dynamical systems approach enables us to avoid this step, or, at least, to make the idea of a system's self-description less mysterious. Once again a classical mechanical example is probably the intuitively clearest model to employ.

Imagine a set of particles moving under their mutual interactions. Perturbation of the motion, most particularly displacement of a single particle, may create forces tending to restore the system to its original state of motion. Naturally the action is reciprocal, but in a large system the individual particle's effect on the totality will be negligible. This will appear quite simply as the particle being constrained by the rest of the system. It is then clear that what we have called coherence is a particular form of dynamical stability.

To understand how this helps to clarify Pattee's position we need a few ideas about stability, so it is worth making a digression to introduce these. Recall that a dynamical system is, in effect, a set of first-order differential equations. The collection of dependent variables, say n in number, are quantities characterizing the system's state, and the independent variable is, in our examples, time.

Geometrically we can envisage the dependent variables as co-ordinates in an n-dimensional space. A state is then represented by a point in this space and the dynamical unfolding of the system's progress with passage of time by a curve showing the position at any time.

A sufficiently good aid to visualization of the concepts described here can be achieved by concentrating on a three-dimensional example, say a three-species ecosystem, thought of as a triple of axes in the ordinary physical space we inhabit. Bear in mind, though, that our principal concern is with complex systems in which n is large.

Assume now that there is a differential equation for each variable which gives its rate of change as a function of the current position. So the set of all the equations tells how the state-point will move from any given point. Following the succession of positions as time passes generates the orbit through an initial point. Having an explicit description of all orbits is mathematically equivalent to solving the differential equations.

In this pictorial representation, an externally imposed constraint is a declaration that the system variables are not really independent, but tied together through stated relationships. So the position of the system cannot be freely specified but is forced to lie in some subset of points, say a curve or surface in the state-space.

Mathematically, this is describable in two ways. Either we retain the original variables, adding to the equations extra terms which describe forces responsible for restricting the motion; or we re-vamp the description in terms of a smaller set of new variables which serve to fix position within the constraint subset.

Returning now to the unconstrained case, consider a single orbit. If we let time run indefinitely, to where does the orbit go? This asymptotic limit, as time tends to infinity, might be a single point, a curve or something of higher dimension; the orbit might even wander around indefinitely without settling down to a simple limit. Let's assume that this limit, the so-called attractor of the orbit, does in fact have a simple character; furthermore, imagine each orbit to have such a simple attractor, and that wherever the system starts it goes asymptotically to one of just a few such ultimate subsets.

These attractors can then appear like constraints. Particularly if the time-scale for the dynamical events is short compared with that employed by an observer of the system, the state will appear to jump rapidly on to the attractor and stay there. We need only assume that all the usual initial conditions have orbits going to the same attractor and the observer will see an apparent self-organization of the system to a coherent behaviour, admitting a simple description in terms of just a few aggregate variables.

These variables can approximately describe the actual position when the system's state is near the attractor. Pattee's rather mysterious internal self-description may then be considered as a projection of trajectories close to an attractor on to the attractor itself.

It is worth re-emphasizing that this kind of discussion can as well apply to perturbations of ecosystem variables, neural network firing probabilities or any situation well modelled as a dynamical system.

Dichotomies or dialectics

Let's look again at the reductionism/holism dichotomy. The most direct and immediate answer to reductionism is to insist that a complex entity is always more than just its parts. That which is added to those parts is the structure of their behaviour as a coherent whole.

In the light of the mechanical materialist philosophy still pervading the natural sciences, this seems a feeble rejoinder to the reductionist claim. "Structure" appears somewhat vague: an ephemeral wraith, when contrasted with the hard, gritty reality of the matter which is structured.

To the extent that this feeling prevails, I would maintain that the information revolution instituted by cybernetics has not yet been acknowledged. The lesson we must still grasp is that the way matter is put together – whether we call it structure, order, organization or what have you – is as material in its causal effects as matter itself.

It might even be said that objects and relations between objects have identical ontological status. Ollman,[3] in fact, develops a purely relational philosophy which I find very appealing. However, a full working out of such an account would, presumably, need a purely relational logic with no atomic entities and I do not know if this is feasible. Moreover, intrinsic properties of objects, such as the mass m in our first example, would need re-interpreting in relational terms. Such a goal, illustrated by Einstein's early concern with Mach's principle, has been a perennial temptation to theorists, but has never been carried through successfully.

The reductionist response to a claim for the independent status of organization is to lean pretty heavily on the idea of self-organization. For if the nature of the parts of a complex system determines that they fall into an autonomously organized form, then it can hardly be said that the organization is something extra. It arises automatically.

This is, I think, the most serious objection to holism and should make one pause before dismissing the reductionist programme in any particular instance too readily. In the end, though, this objection can only lead to a quantitative revision of the role of organization as something over and above the components. From the discussion of autonomous coherence in the previous section, something extra must always be inserted into the most spontaneously self-organizing system; namely, a setting of the initial conditions sufficiently close to the appropriate attractor.

It's worth noting that, while the work of Prigogine *et al.*[4] affords other schemes for self-organization and gives fresh hopes for modelling organized systems in a reductionist way, it doesn't essentially affect our present discussion. For it requires that the system – open now – be born into an appropriate environment affording necessary material and energy flows. This merely pushes the problem out into a wider, already organized super-system and still demands appropriate initial conditions.

So all reductionist programmes presuppose a context with some degree of external organization. Once this is granted holism and reductionism may be seen as compatible because they attack different aspects of a system. The latter presupposes context, expressed through parameters, initial conditions and the like, while the former is concerned with those features as the flexible, and interesting, determinants of the system's behaviour. Reductionism, as here characterized, is concerned with possibilities for autonomous coherence of conceptually closable systems, whereas holism, focusing on the openness, emphasizes the control-from-above features of a command hierarchy.

This rather involved delineation of the dichotomy makes it suspect to clearly map the two sides into moral categories, such as progressive or reactionary. As a social example, consider the electronic battlefield with fail-safe computer regulation and input/output access to a few top button-pushers. I'd certainly have no hesitation in declaring this morally vile, but whether the concomitant view of (i) the computer system and (ii) human society is to be described as reductionist or holist is not at all clear. This would depend on which of the two systems was under discussion, with the other considered as context.

A number of the arguments offered up till now become particularly transparent if we think about a simple device of great conceptual generality, namely a switch. This consists of two stable states for its internal dynamics which are separated by a barrier of height sufficient to forbid random switching through thermal or other noise. It is an open system with switching between the two stable positions being effected by appropriate inputs.

For concreteness, we may imagine a mechanical switch with a pivoting contact arm held in either switch position by springs. More abstractly this can be conceived as a potential energy junction $V(x)$ with two minima. An externally applied constant force F contributes an extra potential $-Fx$ and the graph of the total $V(x)-Fx$ is, for positive F, tilted upwards to the left. For sufficiently large F the left-hand minimum is eliminated and the internal dynamics carries the system rapidly from any given initial position to the remaining stable point: conversely for negative F.

Apart from mechanical examples, qualitatively similar behaviour can be manifested in diverse situations, ranging from mutually inhibiting neuronal pools to magnetization orientation. In all such devices we find a neat example of levels of description. The internal dynamic which determines the switching takes the external agency for granted. For the user, however, the switch is just . . . a switch; the exact mechanism can be ignored.

The possibility of two accounts relates to the existence of two time-scales. When we referred above to the *constancy* of F and to the *rapidity* of the internal state changes, this was a simplification of the fact that externally imposed variations in F must, if the switch is to behave at all like a switch, be slow relative to internal changes.

A peculiar consequence of this two-level structure is the irreversibility of the upper level. Even if the internal dynamics is theoretically reversible, say Newtonian mechanics, a particular switch position retains no record of earlier switch positions. In the sense that this history of former switch positions cannot be reconstructed, even if the exact history of inputs is known, a switch can be called memoryless.

In case it is thought that a switch is too trivial a concept to carry much of a philosophical burden, it's worth remarking that any finite state machine is decomposable into reversible machines, whose history *is* reconstructable, and two-state switches.[5] This means, essentially, that any well-defined model of reality contains history-dependent and history-independent aspects.

In the social sphere, this remark carries particular force when we note that the idea of a switch is general enough to include any kind of facility for permanent registration of information: writing, film, artefacts etc. I believe this sheds some light on the synchronic/diachronic distinction, for it affords the possibility of a further kind of conceptual closure, permitting us to sever an object from its antecedents. If any analogies between holism/reductionism and synchrony/diachrony are thereby permissible, it is amusing to note that holism would be related to structuralism whereas historical materialism would be akin to reductionism.

So much for current vogues. I think, though, I'd rather emphasize that a properly dialectical philosophy should refuse to jump down on one side of these dichotomies, but, rather, hold both sides simultaneously in mind.

Matter and information

The distinction between a system and its state is at the heart of the matter/information distinction, a fact which becomes particularly clear in our switch example, where the choice of state can amount to one bit of information. Actually "information" has come to mean a quantitative measure of particular kinds of organization, but we will here use the word in a looser sense – that the state of a system is "informed" by its surroundings without implying quantification.

I intend it to be understood as identical with that much abused term "reflection". A key concept of epistemology, this term has been grossly caricatured by its critics as a concept of the brain acting like a mirror of the external world. (Maybe such critics should take the complex physical processes involved in mirror reflection less for granted – and more reflectively!)

I will also use the terms "matter" and "energy" interchangeably. This will be adequate for our discussion, though the informing of the two is, possibly, the basis of the distinction between "information structure" and "circulating information" so fruitfully deployed by Laborit.[6]

The cybernetic introduction of self-regulation and goal-direction into the theory of deterministic systems opened the way for an understanding of living and conscious systems in the framework of materialist philosophy. Once attention was directed to the specific structure which a system must possess to achieve stable, purposive behaviour, the simply quantitative notion of energy became inadequate. More qualitative questions concerning the source and destination of the energy within a system – the structuring of energy therefore – came on to the agenda.

Energy may be directed along physical channels such as cables, wave-guides, pipelines and so on, or be generally disseminated to be picked up by receivers tuned to suitable characteristics. In either case a relationship is established between emitter and receiver: energy is structured and we can conceptually sever the quantity of energy from its patterning. General thermodynamic arguments indicate a limitation to this separation, there being a theoretical lower bound to the energy package required to carry one bit.[7] At normal temperatures, though, this magnitude is so small – of order 10^{-20} Joules – that the conceptual distinction is fruitful.

Unless an adequate, purely relational ontology can be worked out, this puts dualism back on the agenda in a materialist way. If mind is not an object but a complex set of relations between objects, there is no reason in principle why it should not be isomorphically mapped into some new, sufficiently complex material system. However, it could never be floated free of a material base, for while matter and information are conceptually distinct, they are inevitably inter-related. This is because (i) the role of information transfer is the triggering of energy release in a receiver and (ii) the storage or transmission of information requires a material or energetic carrier.

The most striking difference between energy and information is that the latter does not satisfy a conservation law. If an adequate energy supply is presupposed, information can be copied without limit. This is manifest in innumerable cases ranging from xeroxing to the spread of epidemics, from photography to action potential branching along axon collaterals.

Failure to pay explicit attention to this difference traps the terminology of social sciences in the energy metaphor of nineteenth-century science. For example, psychoanalysis is grounded in explicitly energetic descriptions of mental events, stemming from Freud's original Q- concept. This has influenced even the most progressive of contemporary humanistic psychologists, and it is almost a cliché for people to speak of "lacking energy", of "feeling blocked" or of "getting energy from others" and so on. A useful attempt to reinterpret psychoanalysis using cybernetic ideas has been made, however, by Pribram and Gill[8] and a tremendously sustained and powerful attack on a vast range of such problems is to be found in Wilden.[9]

A most crucial task is to re-vamp standard accounts of political economy, to free them from an undue reliance on energy-like concepts such as "labour power". I feel that the creation of surplus value is simply incomprehensible without ideas relating to information. For the task of the worker is primarily the shaping or informing of matter. From his or her own local standpoint in our technologically sophisticated society, the provision of energy to the machines which he or she uses is a pre-given fact of the production process.

This is being highlighted through the burgeoning of computer-based production. Differential costs of software as against hardware are rising at a phenomenal rate. The machine is, for many conceptual purposes, just a fixed aggregate of switches and the programming problems are about how to sequence their states into suitable configurations. Yet still it is difficult to think of this sort of information injection as a kind of primary production and its role in the creation of value remains inadequately theorized.

Information and history

In recent years there has been an increasing interest in the production of entities other than commodities. Most striking have been discussions of the production of knowledge leading, in particular, to a radical critique of the status of science. Furthermore, there has been a strong current emerging in psychology to demand an examination of the social production of the individual psyche. These trends form the centre of emphasis for the general critical attitude – that no aspect of the human condition is pre-given; all are constructed within definite social formations.

However, there have been moves beyond this position to what I see as a kind of social solipsism – an argument that scientific knowledge is nothing but a shared code for conceptually structuring the world which is specific to our current society. To parody a famous rejoinder to similar attitudes, "the sophism of such philosophy consists in the fact that it regards science as being not the connection between society and the external world, but a fence, a wall, separating society from the external world." Indeed, it is then an inevitable step to dismiss an external pre-given or natural world altogether, regarding it as a social construction.

The underlying fallacy is that of pushing the parallels between commodities and knowledge too far. Not sufficient attention is paid to their different modes of *consumption*. The feminist critique of the family drew attention to the need to understand the whole cycle of production and consumption in a systemic way. It is no longer legitimate to restrict analysis of social contradictions to the major sites of commodity production. An important component of production, the worker as embodiment of skills, is itself a commodity produced in particular circumstances. The purchasing power of the wage can be dually seen as the mediator of commodity consumption and of labour-power production.

Now it is evident that there is no exact counterpart in a circulation of knowledge. Certainly the worker is constructed not just through commodity consumption but through informing with particular skills. Although the latter is referred to as consumption of knowledge, there is a crucial difference; commodities are destroyed by consumption, knowledge is not.

This is not trivial. Knowledge is difficult to destroy because it is an indefinitely reproducible information-structure of society. A consequence of this permanence is the tying together of temporal events, which enables us to break free of the social solipsism described above.

This point is worth making in some detail. In individual perception, solipsism is unavoidable, since a hypothetical isolated individual has no way to factor out from its interaction with the world those changes which result from its own manipulations and those which are intrinsic to an object. Naturally this kind of individual is a myth, and every person's perception is socially mediated from the outset. The necessary inter-personal communication this entails destroys solipsism from the start, since we thereby verify that an object's properties are independent of perpetual standpoint.

With consideration of socially produced knowledge we run up against the same question at a higher level. We are all caught up in the specific means of production of our society, and within its own terms it might be impossible to differentiate between truth about an objective world and our constructions. However, not only can we examine the world view of other contemporary social formations. More importantly, we have messages from the past through information carried in artefacts, writings etc. Moreover, each individual endures for a span of time as a material carrier of structure, often persisting through vast upheavals of the social order.

This suggests a definition of the natural/social distinction which is not rigid but is compatible with commonsense meanings. The very way in which the category "nature" is employed characterizes it as those aspects of human experience which are relatively independent of the social formation.

How is this to be understood if all knowledge is produced within definite social conditions? Only through the fact that a change in the social order, no matter how catastrophic, is never totally discontinuous. So we can envisage history as a sequence of temporal cross-sections bound together by the day-to-day continuities of thought and language, embodied in the people constituting the world at any time. Against such continuities, which are themselves supported by yet more permanent material things – buildings, machinery, earth and sky – we can then see the movement of the social order which, in a real sense, re-makes the people, their perceptions, their understandings in a periodic yet evolving succession.

If it is granted that such permanences (or, at least, long-term invariances) exist, then it is hardly surprising that we accord them a different ontological status. So we conceive them as objects standing prior to and independent of the social order.

Conclusion

I've attempted to argue for a relational view of the world through the centrality of the concept of information. This seems to me the only way to appreciate the world as an interconnected collection of levels without falling for interactionism. There is a common flavour to ideas of Gregory Bateson[10] concerning knowledge as an active ecology of ideas in circulation, the notion of reflection, the philosophy of relations of Ollman[3] and many of the currently most interesting topics in the natural sciences, ranging from Thom's conception of resonance between dynamical systems[11] to the theory of software complexity.[12] Each is interested in general relational properties of material systems which find their greatest elaboration in the social world.

However, at the end of the previous section I'd written myself into a tight corner. For the definition of "natural" includes those features of human life, such as war and patriarchy, which have exhibited great resistance to changing with the social order. But what's in a word? I could have opted for a less loaded term, but its shock value is to remind us that, for all our pretensions to social concern, we've not come within half a million miles of knowing the social conditions which would eliminate the chauvinisms we all know and loathe (at least with the top ends of our brains).

Built into the definition is the presumption that all human experience is mutable – only some things are more changeable than others. So the usual disturbing equation of natural with inevitable is not implied. Indeed, human biology might be redefined as the study of relatively intractable human behaviour. Then we see that the really big problems of biology are how to break us out from seemingly fixed behaviour patterns – how to enlarge areas of freedom, no less. The demand for such a programme in biology, with particular reference to sexism, has often been voiced by feminists (e.g. Firestone[13]) and should not be too lightly dismissed.

I should like to acknowledge a debt to Graham Clarke, who has discussed relationships between systems and dialectics with me over many years.

References

1. R. Rosen, *Dynamical System Theory in Biology*, London 1970.

2. H. Pattee, "The Nature of Hierarchical Control in Living Matter", in R. Rosen (ed.), *Foundations of Mathematical Biology*, London 1972.

3. B. Ollman, *Alienation*, Cambridge 1977.

4. G. Nicholis and I. Prigogine, *Self-Organization in Non-Equilibrium Systems*, London 1977.

5. M. A. Arbib, *Theories of Abstract Automata*, Englewood Cliffs, N.Y. 1977.

6. H. Laborit, *Decoding the Human Message*, London 1977.

7. R. W. Keyes and R. Landauer, "Minimal Energy Dissipation in Logic", in *IBM Journ. Res. and Dev*, no. 14, 1970, pp. 152-57.

8. K. Pribram and M. Gill, *Freud's Project Re-assessed*, London 1976.

9. A. Wilden, *System and Structure*, London 1977

10. G. Bateson, *Steps to an Ecology of Mind*, London 1973.

11. R. Thom, *Structural Stability and Morphogenesis*, New York 1975.

12. M. Machtey and P. Young, *An Introduction to the General Theory of Algorithms*, New York 1978.

13. S. Firestone, *The Dialectic of Sex*, London 1971.

9

Matter, Information
and their Interaction
in Memory Processes
Lauro Galzigna

Summary

(1) The need for methods of studying memory processes which do not suffer from the limitations of a *deterministic approach* is particularly strong in the field of biological memory. The *co-operative model* seems to be one of the most promising approaches to a formal description of the non-linear phenomena which characterize biological acivities in general and memory in iarticular,

(2) A *statistical treatment* of the behaviour of systems in which the components show co-operative interactions, allows us to cover a wide array of phenomena governed by types of law ranging from sigmoidal to all-or-none. Co-operative behaviour is typical of *dissipative structures* capable of storing information obtained in their past.

(3) The interaction between *information and matter* can be described in general at a *phenomenological level,* taking into account the influence of structure-bound information on the capability of any material to act as information-receiver and information-recorder.

(4) The relationship between matter and information may be represented as a *dialectical process* and *memory* can, indeed, be considered as a product of the history of any piece of matter organized on a co-operative basis. This type of definition certainly seems self-consistent but is somehow closed upon itself unless the antithetic terms are redefined in the case of highly complex material systems such as brain. In such an instance, in fact, memory is an expression of the interaction between environment-bound information and a structural organization which is, macroscopically, the result of evolutionary information.

Introduction

The basic problem in the description of any memory process, and

136

therefore of that underlying learning, is the definition of the elementary act which, for lack of a more precise term, we will refer to as interaction between matter and information. Such an elementary act could represent the initial event of all processes in which a memory trace is established, among them the complex brain function we call learning. A series of consecutive black boxes seems, however, to bar the road to knowledge of how different levels of integration are linked together; hence any aspect of brain function still retains a large element of mystery. What appears obvious is that memory in general and biological memory in particular are non-linear features of systems of different complexity, in which the non-linearity increases with the increase of complexity of the memory-endowed system.

In the following sections, memory is intended to mean a modification induced by a succession of inputs to a material system, so that the succession is represented in any output of the system. In other words, material systems capable of memory are concrete representations of both space and time; for the latter, it seems that only a non-deterministic approach can explain the relevant mechanism.

Limits of a deterministic approach

The most complete formulation of the causality principle in the post-classical age was advanced by Spinoza. He claimed that, given a cause, its effect follows *necessarily* and no final cause, such as that advocated by Leibniz, has to be searched for in nature.

Causality, elaborated as a philosophical system by Newton, then contributed to by Hume and Mill, finally found, in the hands of Laplace, its most satisfactory utilization and systematization.

In 1927, the year of birth of quantum-mechanics, a sort of collapse of the deterministic approach to the physical world resulted from the discovery of uncertainty. After that discovery the causal order of the world was considered as mere appearance, due to the enormous number of elementary processes which are involved and their influence in levelling out the elementary uncertainty. While this happened to microphysics, for biology in general and neurobiology in particular the Newtonian background remained overwhelming and a sort of causal obsession in neurobiological thinking was the result.

Modern physics is based on either deterministic or indeterministic, statistico-objective descriptions, depending on the size of the natural system considered and on the attention given to either its microscopic components or macroscopic behaviour, whereas deterministic concepts are still prevalent in biology. The limits of application of causality concepts in biology and the detrimental aspects of their use in molecular biology have been discussed in detail elsewhere.[1]

The existence, in nature, of reversible and irreversible processes is, according to Planck,[2] a justification for the use of both dynamical, strictly causal, and statistical methods. This is also a result of the two different aspects of natural phenomena, microscopic and macroscopic respectively. Macroscopic observations produce statistical laws based on chance and probability, while microscopic observations produce dynamical laws based on causality and necessity.

The deterministic attitude of the biologists co-exists with the fact that the majority of physiological laws are statistical and it is not always possible to discriminate, within them, causal connections and simple apparent regularities in the succession of events. As a matter of fact, according to Reichenbach[3] the so-called causality principle consists of a whole set of principles.

Causality is the possibility of predicting *unknown* events on the basis of *known* ones and, in such a sense, it is an inductive procedure, where induction means *functional relationship*. The causality principles are operating in any succession of events which are time-independent and defined by regulative functions. A probabilistic concept of causality has even been presented, and this expands the possibility of a deterministic approach to macroscopic phenomena as well. The two types of law, dynamical and statistical, may therefore be regarded as particular forms of a more general, unified principle of probabilistic description of the physical world. The fact that classical mechanics is a limiting case of quantum-mechanics is an expression of such a unified approach. The common way of posing the concept of causality contains, on the other hand, an animistic element which can be eliminated only through a relativistic reformulation.[4] Indeed, even if a causal link is a connection or correlation of phenomena by which an event B follows every time that event A occurs in a system, such a relationship does not consist of these events alone. The whole background of the system is a vital part of the concept, and if the system, including its past history, changes, the relationship between A and B may change completely.

Co-operative models

When a system can be described by linear equations, the causal series in it which propagate with time have *additive* effects. Systems which cannot be described by linear equations are not characterized by causal connections with additive effects, and very often such systems are described by means of so-called co-operative models. Simple causal relationships cannot be used to describe complex biological phenomena, such as those involving memory, and biological memory is one of the best candidates for description in terms of a co-operative model.[5] Co-operativity has been defined as a way in which the components of a system act together so as to switch from one stable state to another.[6] This definition does not say too much about the detailed nature of phenomena in which the function of the whole becomes something more than the sum of the functions of its parts. Phenomena of this nature are often characterized by sigmoidal relationships between an input response in systems composed of elementary components, each of which can exist in either of two states.

The sigmoidal function is not necessarily limited to co-operative phenomena, but several complex natural phenomena are recognized as co-operative because non-linear output follows from a linear input and the experimental expression of such a correlation is a sigmoidal relationship.

The dissociation curve of oxyhaemoglobin is a typical example of a sigmoidal function: the function expresses oxyhaemoglobin formation (output parameter) in terms of oxygen partial pressure (input parameter). An empirical expression has been recently elaborated[7] to fit, in an easily computerizable fashion, a number of experimental curves of sigmoidal type. This expression is:

$$x'(t) = \exp\left[4 \int x(t)/t\, dt\right] - 1 = \sum_{i=1}^{4} A_i x^i$$

where A represents constant values, t is the input parameter and x the output parameter.

A statistical-mechanical description of co-operative systems is possible, so that a phase transition can be represented by a function whose sigmoidicity increases with increasing number of components in the system. What is referred to as a phase transition is a way of indicating the "qualitative jump", the sudden change from one state to another with changed symmetry or the catastrophic setting-up of a new order in a system that has apparently forgotten its past and broken a causal chain in the relationship between input and output.

Consider a system whose components have two possible states $(+)$ and $(-)$, with a^+ and a^- the numbers of components in the two states. Let $r = a^+/a^-$ and $\theta = a^+/(a^+ + a^-)$, the fraction of components in the $(+)$ state.

When only four independent components characterize the system, as in the case with a tetrameric molecule with binding sites which can be either occupied or unoccupied, the partition function is:

$$Q = (1 + r)^4 = 1 + 4r + 6r^2 + 4r^3 + r^4$$

Co-operative transitions may occur with no intermediate state, when a shift of one component from $(-)$ to $(+)$ forces a shift of all other components to $(+)$. In that case $\theta = r^4/(1+r^4)$ or, in general, for a system with N components $\theta = r^N/(1+r^N)$ A plot of θ against the input necessary to effect a transition is sigmoidal and the sigmoidicity increases with increase in the number of components. As $N \to \infty$ a true phase transition occurs and an all-or-none-type law is observed.

Statistical treatment

The phenomenological equation proposed[8] to describe co-operative processes involving formation (F) and removal (R) of an information vector is:

$$\frac{dx_i}{dt} = (F_i - R_i)x_i + \sum \phi_i x_i$$

where the term $_i o_i x_i$ is the sum of fluctuations around equilibrium and environmental effects. The expression is a typical "master equation" of a process depending on both chance and necessity. When formation equals removal the expression reduces to:

$$\frac{dx_i}{dt} = \sum_i \phi_i x_i$$

and after development as a Taylor-McLaurin series and integration we obtain an expression similar to that for θ above, that is, a sigmoidal function.

It is known that in the neighbourhood of a critical point of the input parameters, systems are very sensitive to external perturbations. Two main independent parameters regulate critical phenomena: an order parameter and a dimensional parameter, both related to macroscopic characteristics. In other words, phase transitions depend on the number of dimensions of a system and on either one (Ising's system), two (XY model) or more than two order parameters.[9] In the last case, which is the most general, a physical tool called *renormalization group theory* can be used for calculating the critical values.

Near to a critical point, a system hesitates between order and disorder, and this is shown by fluctuations. Very different types of transition can be described by the same law once the Hamiltonian expressing the energy balance is formulated.

Tools such as the renormalization group can be considered, at least in principle, as applicable to biological systems, although these present enormous difficulties for any type of formalization.

The use of concepts such as that of dissipative structure seems, on the contrary, less difficult.

Dissipative structures

Co-operative transitions resulting in a change from one phase to another usually involve a switch in the state of a system from disorder to order. The process often has features of self-organization or, at least, amplification of an effect in the space of the system.

Self-organization is a process that can occur only in open systems constantly interacting with the environment under non-equilibrium conditions. For other types of system, e.g. closed systems, entropy fluxes from the environment or within the system would be zero. A strong thermodynamic coupling of the components of a system is another condition for self-organization. In living systems we have the co-existence of ordinary structures, held together by molecular forces and subjected to classical equilibrium thermodynamics, and dissipative structures, held together by a dissipation of energy and following non-equilibrium thermodynamics. Physical dissipative structures, such as those appearing in Benard's phenomenon, have been recently studied [10] as examples of production of order from disorder. Biological structures as well as physical dissipative structures are characterized by a high degree of functional order: the failure to follow the second principle of thermodynamics is a feature common to both physical and biological "open" systems, capable of exchanging matter, energy and information with their surroundings. An order principle completely different from the order principle of Boltzmann is valid for such systems and is called *order through fluctuations*. [10]

Creation of order is common in conditions far from equilibrium in the same sense in which order destruction is common in conditions close to equilibrium. Far from equilibrium we can, in fact, have an amplification of fluctuations and stabilization of such amplification in a system due to a flux of matter and energy from the surrounding environment. The type of response of a system to any form of input is deterministic until a critical point is reached: the further evolution beyond the critical point is statistical. The boundary conditions and the non-equilibrium constraints between the system and its environment are crucial in establishing the position of the critical point.

A description of a phase transition in a dissipative structure has been accomplished[10] by using the concept of "nucleation". When we have a fluctuation in a certain region of a system, the interaction of the region with the rest of the system tends to quench the fluctuation. The dimensions of the fluctuating region, depending upon the dimensions of the input, regulate the possible amplification of the effect. Only fluctuations beyond a certain value will resist the quenching influence. Far from equilibrium, the activity of each component of a system seems able to escape the levelling effect of averaging. The competition between dissipative processes, which tend to reinforce the fluctuation, and the exchanges with the environment, which tend to quench it, will determine whether the fluctuation invades the system or disappears.

Information and matter

Any physical process which exhibits hysteresis is an example of storage of information carried by a material support. This occurs when an interaction has taken place between an input of information and a support consisting of a certain type of matter.

The process of information storage carried out by the human brain may include an initial physical step interconnected with other processes and concurring in the complex brain activity called memory.[5]

Brain function in general depends on transformations of information coming from the outside world and interacting with brain structures. It is very difficult, at the moment, to differentiate, in the context of neural activities, signals from noise and to clarify, in terms of information theory, what happens in a system constantly modified by growth and metabolism.[11]

The central nervous system is constructed so that it can receive the environmental information (chaotic as a whole) in a fashion strictly dependent on the structure of the receptors, as an ordered and coherent flux.

In other words, the central nervous system abstracts from the incoming information only the part that contains some regularities; the rest is ignored. The world appears to us quite comprehensible and ordered because these are the characteristics of our perceiving structures, the latter being an end product of evolution and natural selection.[12]

Information theory originated from consideration of an abstract communication system in which a message is transmitted through a communication link or channel, minimizing the noise produced by the channel itself.

If the elements $j(1,2,3.....n)$ correspond to questions and $i(1,2,3...m)$ are the answers, in a communication system equipped with a scanning device, the amount of transmitted information I for a receiver, after having read a message, is

$$I = -(H_{after} - H_{before}),$$

where H is the so-called "confusion functional" expressing the mean information per position in the message and given by

$$H = -\sum_i p_i \, ld \, p_i,$$

where the p_i are the probabilities of the answers.

When the probability assessment is not changed after the message (because it tells what the receiver already knows) we have redundancy. If the message contains signals that do not correspond to any element of the receiver's memory, the answer refers to an unasked question and the information is irrelevant. Each message has a certain information content with a quantitative value and is completely independent of the meaning of the message or its semantic value.

It is important to distinguish between free- and bound-information: free-information is the type of information of an abstract nature, such as that contained in a numerical sequence with no particular physical meaning; bound-information is the type of information contained in a physical system to specify its structure, and this is the only type of information related to entropy. The possible "cases" of bound-information can be identified, for instance, with the microstates of a system, and the relationship between statistical entropy or disorder and information which reduces it is intuitive.[13]

Another common distinction which is important in the present context is that between micro- and macro-information. Microinformation is that linked to the individual degrees of freedom of the matter constituting an object and, in terms of the memory capacity of the object, each element of memory corresponds to a degree of freedom and each state of a degree of freedom corresponds to a state of memory.

Macro-information is that of macroscopically readable signals, each of which consists of many individual micro-informations. The biological information of the genetic code has characteristics intermediate between macro- and micro-information.

To quantify information in general we need to consider a problem (P), *well defined* by a number of possible answers X, and the probability p that one of the answers is the right one. Our incertitude is given by a function: $S(P/X)$.

A new message concerning the problem gives a new element X' of X, so we can consider the difference $I = S(P/X) - S(P/X')$ as the received information. Information can therefore be regarded as the difference between two incertitudes or entropies, each one given by $S = -kp_i \ln p_i$, where k is the Boltzmann constant and p is the probability of occupation of a microstate.

All this imposes several restraints on the information coming from the external world and impinging upon the central nervous system. The act of "posing a problem (P)" and "defining a set of answers (X)" corresponds to questioning the world in a fashion determined by the structure of the questioning apparatus. *The structure-bound information will therefore determine the ability of any material to act as an information-receiving device.*

Phenomenology of matter-information interaction

We shall try to isolate here the mechanism of information storage, the act of receiving and recording information by any material capable of memorizing the incoming signals. This does not amount to reducing the process of memory to that act, but to understanding the meaning of such a general memory pre-requisite in material terms.

Information is a physical quantity, always requiring for its transport a material carrier that we can call an information "support" (e.g. light, sound, etc.). All material bodies, on the other hand, are associated with a certain amount of information, which is related to the extent of order within the body (or to what we can call the extent of ordered complexity of the body).

In any case information must be considered as something *different* from matter. Matter cannot be created, but a process of self-organization of matter can create information. Material bodies can exchange information with their surroundings and some law of symmetry should probably be taken into account for a proper description of such an exchange.

We point out here some analogies between the interaction of matter with information and other natural processes of interaction between different forms of matter, such as the interaction between matter and radiation. The interaction between matter and electromagnetic radiation is characterized by a number of phenomena such as polarization, absorption, reflection, diffusion etc. They all depend on the fact that matter is made up of charged particles in motion and the electromagnetic radiation impinging upon matter must perturb the electromagnetic system *within* the matter itself.

Matter, on the other hand, if suitably excited, can become a source of electromagnetic radiation. The velocity of electromagnetic radiation which crosses a material body depends upon a characteristic of the body, its refraction index given by $n = c/c'$, where c is the velocity of the radiation in the vacuum and c' its velocity within the material.

The quantum structure of matter implies that only certain transitions are *allowed*; and a further restriction which regulates the interaction between matter and radiation is the existence of *selection rules*. The rules of allowed transitions and selection which regulate quantum-mechanical phenomena have their analogues in living matter if one considers the transmission of heredity, or evolutionary information. It is quite likely that, in due time, similar analogues will be recognized in the elementary mechanisms for recording environmental information, which is one of the basic attributes of living matter.

The use of selection rules, for the moment, has allowed the formulation of a theory of evolution based on the self-organization properties of matter.[8] Selection means, in such a context, the existence of constraints which imply the maintenance of systems far from their thermodynamic equilibrium. In such conditions, as we have seen in the case of dissipative structures, fluctuations can bring the system towards a new steady state and can create order, that is, organization.

Dialectical approach

Perception is a phenomenon made possible by the interaction of a limited portion of electromagnetic radiation, visible light, and material structures such as the receptors of the retina.

The wavelength range suits the function of visible light as a support for information concerning macroscopic structures, which appear well defined and limited because of the resolving power of visible light itself. With hypothetical X-ray-sensitive receptors, the boundaries of objects would appear undefined and shaded. Information is the factor which differentiates being from appearing and, in a sense, the interaction between matter and information can be regarded as a dialectical process whose product is memory. The dialectical definition puts due emphasis on the role of history in defining the actual characteristics of any system capable of memorization. Somewhat underestimated in such a definition, however, is the role of the structure and therefore that of the bound-information, which is the limiting factor of any recording.

The limitations of the dialectical approach are perhaps due to the residues of idealism left within dialectics by some philosophers, particularly Hegel.

Hegel's principle of contradiction tried, indeed, to reconcile empiricism with idealism by means of the dialectical link representing a synthesis of thesis and antithesis in a unit formed by being and its opposite. The concept of becoming, abstract in Hegel, evolves into that of history in the Marxian elaboration of Hegel's philosophy, and this "concrete" reality of history replaces the abstract man of Hegel.[14]

Becoming and history create appearances through their interaction with material supports in a way that resembles what we have described as the interaction between information and matter. It has already been pointed out that information can be interpreted on the basis of either idealistic or materialistic concepts, depending on its thermodynamic or probabilistic formulation, respectively.[15]

The contrast between the phenomenological and dialectical view is based on how consciousness is considered: it is an invariant for the phenomenological thinker, whereas it is a variable for the dialectical thinker. For the latter, the data of the conscious mind aren't known objects but just tools, and the dialectics of knowledge allow one to reach a level of reality which goes beyond the limits of perception.

Dialectical thinking claims its pre-eminence over mechanical materialism since the latter assumes an immutable essence of things. The so-called reflection theory, the most recent formulation of the dialectical theory of knowledge, establishes between things and sensations a relationship similar to photography: sensations are not to be considered in the absolute but must be included in a process of dynamic knowledge. According to modern epistemology, the question is, to what extent a result which is produced by the history of knowledge can act as knowledge.

Memory in brain structures

In the complicated machinery of the brain there are different types of information, different types of memory and different types of integration level. If the concept of space-time is an "a priori", living matter is but a demonstration, an epiphenomenon, a perceived form of the "a priori". In contrast, if matter is taken as an "a priori", space-time is only a perceived quality of matter. The fact that we have no method for verifying concepts deriving from sensorial data, beyond the data itself, puts us in a quandary.

This seems to be the situation if we consider, in our analysis, a material system such as the brain. It is difficult, in this case, to make use of relativistic concepts, because the system coincides with both a space-measuring and a time-measuring device. It is in fact a space-meter, a self-observing observer and a clock, i.e. an expression of the distance of its own non-equilibrium state from its equilibrium state.

Sensorial data are part of that entity which we call our existence, the latter being the process of our living, with its implied accompaniments of seeing, hearing, touching and and communicating with what is surrounding us. The operational capacities of senses can be amplified through instruments, and the fact that senses are transducers of environmental information renders central to the self-definition of existence the elaboration undergone by environmental information within the central nervous system. Concepts originating from such an elaboration are a synthesis of information deriving from the comparison of actual data with memorized data.

The simplest situation is that of imprinting: in this case the recording of data is not apparently influenced by the comparison with previous data stored as memories, but seems to depend only on the interaction between environmental information and cerebral structures.

Environmental information is a type of bound-information since its possible cases correspond to different aspects of a physical system, namely the environment. Cerebral structures are, on the other hand, the expression of another form of bound-information. Both types of bound-information are, like all possible types of bound-information, particular species of free information.

In energy terms information corresponds to negligible amounts of heat, and a factor of 10^{-16} is the transformation coefficient[16] of information units into entropy units. This means that in a system such as the brain, composed of 10^{10}-10^{12} cells and 10^{13}-10^{15} synapses, the entropic changes depending on environmental information can be non-negligible if the recording of information is diffused to all cells or synapses of the system. The consequence of such a consideration is that, under particular conditions, an "organic", material damage may be induced on the brain by certain information, which ultimately acts as a physical injury comparable to that induced by psychoactive drugs.

The structure of a system can generally be said to be organized if its existence is necessary to maintain some functional organization, that is, an organization of the operations of some function. What seems equally important for the recording of information at the macroscopic level, in a system such as the brain, is its actual structure (bound-information) and its status as a result of evolutionary history and environmental effects. As pointed out elsewhere: "The effects of the environment would thus modify the cerebral organization itself, and would not be limited to the release of foreseen stereotyped responses."[16]

If a structure is characterized hierarchically, a reduction of the overall bound-information results. The typical situation encountered in self-organization phenomena would then occur: the type of organization is specified by internal bound-information, while the information coming from the outside world provides the constraints. The type of change representing the recording depends therefore on both forms of information, one of which has mainly the character of necessity, the other of chance. A principle of *transfer of order* has been formulated[17] to couple the acquisition of information and therefore of order, by a so-called organized macroscopic system (say a biological system), to an increase of disorder of previously ordered microscopic components of the system. The increase of entropy accompanying the formation of a new structural organization (i.e. the recording of environmental information) is paid for by the microscopic constituents of the system.

The transcription of changes of electric fields into macromolecular conformation has been shown to occur with an actual increase of macromolecular disorder.[18] This is a microscopic event, basically a transfer of order, appearing at the macroscopic level as increased organization – a learned pattern. The increase of specific information that accompanies the appearance of new structures is related to the increase of entropy parallel to their formation.

Conclusion

Different methods of formalization utilized to describe memory processes in systems of a low integration level – molecular aggregates – have been brought together in this paper. This has been done in an attempt to see whether a theoretical framework may exist into which we might put results obtained in studies of the memory function of complex hierarchized systems, such as the brain.

As recently pointed out,[19] adult learning depends on the activity of neurons capable of plastic modification throughout life. The capacity for plasticity is itself epigenetically determined, but what is important is the way in which the biochemical machinery changes are linked to changes at a higher level.

Out of the big black box that still exists between environmental changes and their transformations into cerebral bound-information, one thing perhaps emerges: different orders of amplification phenomena are linked together in a change of events that, from physical changes resulting from matter-information interaction, evolve into physiological responses. In that chain, progressively increasing orders of co-operativity must be involved, but the general law which regulates such co-operativity, and therefore the progressive "jumps" from lower to higher integration levels, must be unique. Iteration of non-linear functions from computer technology can be used to build up models of such a chain of events.

References

1. B. E. Wright, *Trends in Biochem. Science*, no. 4, 1979, p. 110.

2. M. Planck, *Wege zur physikalischen Erkenntnis*, Leipzig 1954.

3. M. Reichenbach (ed.), *Modern Philosophy of Science*, Los Angeles 1959.

4. P. W. Bridgman, *The Logic of Modern Physics*, New York 1927.

5. L. Galzigna, in H. Baum and J. Gergely (eds.), *Molecular Basis of Medicine*, Oxford 1979.

6. F. O. Schmitt, D. M. Schneider and D. M. Crothers (eds.), *Functional Linkage in Biomolecular Systems*, New York 1975.

7. G. Torresin and L. Galzigna, *Spring Meeting of Soc. of Ind. & Appl. Mathematics*, June 11-13, Toronto 1979.

8. M. Eigen, *Die Naturwissenschaften*, no. 58, 1971, p. 465.

9. K. G. Wilson, *Phys. Rev. B4*, 1979, p. 3174.

10. I. Prigogine, *La nuova alleanza*, Milan 1979.

11. J. von Neumann, *Lectures on Probabilistic Logic*, Pasadena 1952.

12. L. Jacob, *Evoluzione e bricolage*, Turin 1978.

13. L. Brillouin, *La science et la théorie de l'information*, Paris 1959.

14. L. Geymonat, *Storia del pensiero scientifico*, vol. 6, Milan 1972.

15. V. Somenzi, *Quad. crit. marxista*, no. 6, 1972, p. 209.

16. H. Atlan, *L'organisation biologique et la théorie de l'information*, Paris 1972.

17. J. Polonsky, *Ann. de radioélectricité*, no. 17, 1962, p. 227.

18. L. Galzigna and B. Voigt, *Ital. J. Biochem.*, no. 27, 1978, p. 168.

19. S. P. R. Rose, in Y. O. Tsukada and B. W. Agranoff (eds.), *Neurobiology of Learning and Memory* New York 1980, pp. 179-91.

10
Reduction Reassessed
Werner Callebaut

Introduction

After my return from Bressanone, while I was still recovering from
the intense interactions there, I had to explain the results of our meeting to some friends and colleagues. I told them curtly but still euphor-
ically that we had attended reductionism's funeral. Immediately,
some of my philosopher fellows who are always eager to flirt with the
latest brand of anti-scientism they encounter asked whether I really
wanted to imply that atomism, mechanism and the analytic method of
science were passé and had to be abandoned, something which they
had come to realize themselves quite long ago. Rather unexpectedly I
found myself in a delicate position. I had to tell them one must make
the necessary *distinctions*; and here we went on for another couple of
hours, discussing dialectics, epistemological pluralism, the limits of
reductionist strategies in biology and psychology, the heterarchical
organisation of levels of understanding, and what-have-you.

Although I realize that I am no good at story-telling, I hope to have
made clear that there is something paradoxical about one of our conclusions – that *we can accept reductionism as an explanatory stratagem
when it is confined to relating* (in a sense to be specified) *levels of explanation for a particular phenomenon, but that a philosophy of global reductionism is objectionable*[1] – unless we are able (i) to state precisely what is it
we accept and what we reject about reductionism in general and in
particular cases, and why; (ii) to specify an additional (dialectical?)
explanatory "principle" (or several such principles); and (iii) to make
explicit the relationship between (i) and (ii), which have to complete
(and probably also to "correct") each other. The paradox I am referring to does not disappear with Hilary Rose's expression according to
which reductionism has initially been "liberatory" but has now
become "oppressive". For this is of course a simplification that does
not apply automatically. Not all modern reductionist arguments are
necessarily bad, as she herself acknowledges. On the other hand,
certain older forms of reductionist thinking were as oppressive as
today's.

One should read, for instance, Ernst Bloch's penetrating critique of pre-war health and welfare programmes in *Das Prinzip Hoffnung*. Bloch alertly points to the naturalistic trend which pervades modern medicine and which originated with Malthus's population doctrine, a "social utopia in medical soil" which was "projected back" into nature by Darwin. He writes:

> Here, the means for improving health are always located as far down and as deep as possible below the level of the real person and his or her surroundings. So even without the literal presence of Malthus, we have an at worst deliberate or at least unthinking narrowmindedness that believes it can determine the full sickness of people as it were from a drop of blood sent to the laboratory. Our attention is turned away from the whole, living, suffering people, let alone the circumstances in which they find themselves.[2] [Editor's translation].

In a nutshell and without ever so naming it, Bloch acutely recorded the case of reductionism more than thirty years ago! It is interesting to note also that in the same passage Bloch ironically dwells on "petty-bourgeois" uncooked food-freaks, breathing-exercisers and the like, who represent but another facet of the same capitalist ideology, "even though they do not legitimate imperialist wars"!

To resumé my own position: I believe – and here we probably all agree – that the Bressanone meeting has been rather successful in exposing the ideological and political significance of various recent forms of biological determinism as well as their scientific untenability *in detail*.[3] But I also feel that the *general* concepts of reduction and reductionism, which are crucial to our whole endeavour, are in need of further clarification and explication. This task should not be performed for the sake of scholarship, but because *it can help pave the way* – in an essential manner, I believe – *to a full-fledged liberatory alternative* which, in order to be *powerful* enough, should at least to some extent be general.[4] The conference materials already contain a host of valuable suggestions and hints that go in the right direction; but these insights are scattered and ought to be systematized, which is likely to succeed only by means of a collective effort. Therefore, I would suggest that this task should feature as an important point on the agenda of the follow-up conference.

My aim in this paper is much more restricted. I will try to show that when the history of the reductionist "unity of science" program of logical-positivist philosophy of science is put into proper perspective, a number of interesting conclusions arise that may guide us in solving some of the conceptual and theoretical problems we have pointed to (but only in a rather *ad hoc* way) and that we still have to tackle. Such an *epistemological* approach is rather limited in scope, since it focuses essentially on the "semantics" of the reductionist problematic, often to the detriment of "pragmatic" considerations. Now it is fairly obvious that, from the perspective of the critique of ideology, pragmatic factors are by far the most important ones. Judgements about the "fundamentality" of one epistemic level with regard to another which are the hallmark of reductionist thinking,[5] amount to ascribing an absolute ("real") meaning to one level and a derived ("secondary") meaning to another level. Such judgements depend considerably on what Anthony Wilden calls "punctuation"; and punctuation involves "history, responsibility, and power",[6] which are pre-eminently pragmatic notions. Logical positivists, especially the older generations of them, have been notorious for discarding pragmatic considerations as "none of their business".[7] With this limitation in mind I think it is worthwhile to analyse the logical positivists' successive proposals – their various "logical reconstructions" of allegedly real reductionist practice in science – as "informative failures" (Lindley Darden).[8] Their failures may guide us in asking better questions and constructing more adequate conceptual categories that yield a more profound understanding of both the *horizontal* dimension of scientific development (the succession of theories) and the *vertical* dimension of scientific development (unification by means of reductions or other mechanisms of theory replacement). To date and for the purposes at hand, the logical-positivist analysis of reduction might well be "the best we have", to borrow Giovanni Jervis's expression.

Some people may find an epistemological approach to our subject unattractive or even superfluous. To them I would reply that the gap which at present separates "real" science from philosophy and, more particularly, from epistemology (and vice versa), is itself an aspect of the alienated mode of production we are struggling against. In order to be fruitful, even the best informed case studies of reductionist thinking, or auto-investigations by scientists of their own attempts to transcend reductionism, are bound to rely somehow on a kind of *abstraction* and *generalization* from particular phenomena, i.e. exactly what philosophers of science have set themselves to do in the past, failing to succeed because they were estranged from real scientific practice. Thus a real symbiosis of science and epistemology is in order if we are to avoid the pitfalls of abstract "grand theorizing" as well as those of casuistical empiricism. A parallel with the recent history of materialist dialectics may be instructive here: the justified dismissal of stalinist formalism has resulted in an uncompromising disavowal by large parts of the left of all attempts at a genuine theoretical clarification of dialectical logic, while both general and scientific practice abound with examples of dialectical mechanisms at work.[9]

The many meanings of reduction and reductionism

Before turning to the analysis and assessment of the reductionist programme of logical positivism, it seems useful to distinguish between a number of differing meanings of the concept of reduction that currently occur in discussions of reductionism. My strategy will consist of sorting out those usages of the term "reduction" which have a direct bearing on scientific explanations that take the form of "intertheoretic" reductions. This does not imply that I consider the other, more remote meanings of the term to be irrelevant to our undertaking. Quite to the contrary, they should eventually be reintegrated in the frame of reference that will be outlined here. The following list is by no means meant to be exhaustive; I only present it because it helps to create some order in a rather poorly organized field.

(i) Reductionism is often associated with a way of *simplifying the complexity of reality in our accounts of it*. Thus one type of discourse may be labelled as "impoverishing" the "world" (i.e. some part of our environment with which we interact), while another type of discourse will not be considered as impoverishing: cf. Gerry Webster's comparison of the scientific study of chick memory with Proustian memories. The problem here is to gather *why* we are entitled to do so. In our example, one is immediately and intuitively struck by the difference. But very often, this will not be the case; as when two people with widely differing socio-cultural backgrounds are arguing over a novel, say by James Baldwin, and one feels it to be very deep while the other regards it as rather superficial. Note that what concerns us here is not the same as comparing the *relative* simplicity or complexity of two accounts of the same phenomenon,[10] which may be relatively easy provided an adequate measure of simplicity can be found. (Actually, existing theories of simplicity such as Nelson Goodman's can only account for certain limited types of formal simplicity that are conditioned by the type of "primitives" one has chosen. This heavily restricts their effective applicability, as even the most hard-headed of formalists must admit.[11]) What we are really trying to do here is to compare what is "out there" with our discourse(s) about it. We all know, since Kant, that it is unfashionable to predicate the *ding an sich*

from which we are said to be separated – save for the diehards who
continue to believe in Lenin's paradise – by an unbridgeable abyss.
Thus Goodman defends the thesis that the world "has many different
degrees of complexity as it has many different structures; and it has as
many different structures as there are different true ways [scientific or
other – W.C.] of describing it". In a view such as his, predicates such
as simple or complex, or (un)grammatical, or (in)coherent (etc.)
"apply to the world only *obliquely*, through applying to discourse
about the world".[12]

Yet I think the distinction between the *inherent* ("objective") *simplicity or complexity* of a phenomenon and the *simplicity or complexity of
that phenomenon as due to the eye of the beholder* must – and can – be
maintained if the self-denying positions of scepticism and agnosticism
are to be avoided. Without going into detail here, I contend that a
marxist theory of knowledge should rely on a notion of truth that
intrinsically combines requirements of correspondence and of pragmatic adequacy.[13] An "exact" description of a phenomenon will generally be more complex than an approximate, sketchy description of
it. As Herbert Simon put it, *we may regard the inherent complexity of a
phenomenon, say a system's structure, as "defining a lower bound for the
complexities of exact descriptions"*, while "approximate descriptions
may be even less complex than this lower bound".[14] Inherent simplicity thus amounts to "describability" in principle, in adequately simple terms. Note that one may fail to actually discover the simplest
possible description of a phenomenon and be able to characterize it
only in a more complex way.

Two additional remarks on inherent simplicity are in order here.
Many a systems researcher has been struck by the circumstance that if
one defines the complexity of systems in terms of the cardinality of the
set of their elements, or the degree of interdependence among their
components, or their (syntactical) information content in the
Shannon-Wiener sense, or by means of other criteria, actually
observed systems turn out to be dramatically more "simple" than one
would have expected on the basis of, say, elementary combinatorics.
(Of course, this may not preclude these systems from being already
too complex for a human brain to handle!) Thus Simon has shown that
"Fortune smiled upon Kepler and Newton", in the sense that the
scrutability of "their" solar system hinged on its being far less complex than might have been conjectured on the basis of 272 pairwise
interactions (a sun, six planets and about ten visible satellites, which
makes 17 x 16 potential interactions among the elements of this solar

system). Simon essentially invokes evolutionary mechanisms to account for this complexity differential: evolution "prefers hierarchies", and hierarchical systems are relatively simple.[15] This is not the place to evaluate Simon's views concerning the "near-decomposability" of most natural systems – which comes down to a rather peculiar form of "systems-theoretical reductionism" – nor his theory of hierarchies in general, which has intriguing theoretical consequences.[16] But two ramifications of his analysis have direct bearing on our problem.

First, it cannot be excluded a priori that the reason why most of the complex systems about which relatively much is known have rather sparse "incidence matrices" (matrices which indicate the degree of interdependence or interconnectedness of a system's components) may simply lie in the fact that systems lacking this property are inscrutable in principle, because we only "attend to things . . . that are simple enough to yield to analysis."[17] The implications of such an evolutionary view, if correct, for a marxist theory of knowledge, are rather obvious I think.[18] Its elaboration should enable us to qualify and to tone down appropriately the leninist tenet according to which *we will eventually come to know everything through praxis*, by accounting for the ecological constraints the human species is ultimately subject to[19] in its capacity as a knowing species (these constraints remain largely unknown to date, the contrary creeds of certain ecofascists notwithstanding). I will no longer dwell upon this point now, but merely note that the issue of infinite plasticity (cf. bourgeois existentialism and philosophical anthropology) versus human "nature" is far from being resolved.

Secondly, Simon's analysis is instructive in showing that the amount of interaction in a system (which, for the sake of simplicity, I will consider here as the privileged yardstick for measuring complexity, being aware of its severe limitations) may be limited not only by comprehensibility to the beholder – the former case – but also by *comprehensibility to the system itself*. This will especially be the case for evolutionary systems that change their own structure adaptively (Simon himself speaks of "learning systems"; but I do not think the distinction is very relevant here). Their adaptation may require ways of "understanding" their own structure in order to identify the components in which adaptive changes "should" be made, and to characterize the nature of these changes. Thus *semiotic* constraints may impose bounds on a system's self-renewal.[20] Of course, these constraints may themselves be subjected to evolutionary edification.

(ii) This takes us to a second form of reduction which is often referred to in the literature, one that has only incidentally been mentioned at our meeting: the *conversion of one level of communication into another*. This happens, for instance, when it is thought that some "object" language (the *text*) is in itself the "real" signification of the "meta"-statement (the *commentary*); or vice versa, that the commentary is the "real" signification of the text. Wilden notes in this respect that "(i)t is the notion of a theoretically detectable and somehow absolute level of 'real meaning' in what has been traditionally viewed as a relationship of *reflection* ("Widerspiegelung") which has misled so many commentators into the fallacies of reductionism". Any commentary involves a reduction in this sense, but "the reductionist fallacy" differs "in that it confers a privilege on a certain level."[21]

Whether reductionism in this sense will be viewed as an instance of the "explanatory stratagem" we are after obviously depends on one's view on explanation. During the last two decades, traditional accounts of deductive-nomological and statistical explanation – such as Carl Hempel's – have been severely attacked, and a number of alternatives, some of which are rather liberal, have been proposed. I cannot enter this discussion here; but two brief remarks are in order. Both have to do with the traditional view that *we seek explanations for the "unfamiliar"* (obviously a pragmatic notion), and that we do so by accounting for it in terms of something more familiar.[22] On this account, phenomena may be said to be explained "either by comparing them with other, more self-explanatory happenings of the same kind, or by relating them to happenings of some other sort, which are thought to be intrinsically more natural, acceptable, and self-explanatory" (Stephen Toulmin).[23] In this sense of explanation, our second form of reduction(ism) will in certain cases be said to provide a partial or complete explanation (i.e. when the "happenings" involved take the form of levels of communication). A second account of explanation which may also be said to apply here requires only that certain events be "modelled" or "paralleled" by other events, as in Kenneth Craik's theory of explanation which views thought as the "conscious working of a highly complex machine" that models external events.[24] The same cannot be said of other interpretations of explanation, however. (We will have to return to the topic of explanation further on.) The first form of reduction discussed here is sometimes interpreted – rather oddly – in terms of the second; e.g. when the scientific enterprise is viewed as a "continuous comment" on the (metaphorical) "book of nature" (cf. Goodman's point on obliquity).

Note that I have deliberately refrained from referring to *abstraction* as a form of reduction in this context, though the idea of an "object-abstraction spectrum"[25] is sometimes encountered in the literature. Mario Bunge, for instance, has considered a linguistic stratification based on the notion of *extensive* or *semantic referents*.[26] My restraint here has to do with the difficulty or even impossibility of speaking of "degrees of distance" between the empirical world and our models or theories of it in a conclusive way. It will not do to consider a concatenation of meta-levels starting from an object level (which is taken to refer to "real-world" phenomena) and to say that *in general*, the higher the rank of a level on this hierarchy, the more "abstract" it will be. Rather, one will have to distinguish between syntactic, semantic and pragmatic meta-systems and it will appear that within each type, meta-systems may obtain that are *less* abstract than lower-rank systems.[27] (Note also that abstraction ought to be distinguished from generalization.)

(iii) A third meaning of reduction obtains when *a system* (e.g. a living organism, or a human organization, or an artefact)[28] *is said to reduce the information it processes by means of some selection or filtering mechanism.* This is the kind of reduction which Wilden seems to have in mind when he speaks of "the powerful systems of reduction revealed in human perception (gestalts) and in human communication (signs and signifiers)," the lack of which would make "digital knowledge" impossible. Certain brands of systems theory, such as the "grand theory" of the German conservative sociologist Niklas Luhmann, are even entirely grounded on the concept of a *complexity differential* between a system and its environment. Systems are viewed then as relatively stable "islands" of rather low complexity in an overly complex and variable environment.[29] Luhmann's theory has been repeatedly and rightly challenged from the left (notably by Jürgen Habermas) for both political and methodological reasons, but his central intuition, when bared to its rational kernel, can, I believe, be upheld.

(iv) In classical philosophy textbooks, reduction and reductionism are usually associated with the older empiricists' programme of constructing all knowledge, including logic and mathematics, out of "sense data". This view has been most influential in the nineteenth century, as witness the anti-atomism of the Chemical Society which long refused all papers with a theoretical tenor in its journal and the overwhelming majority of the members of which were declared adversaries of atomism because of its "non-empirical" character.[30] In John

Stuart Mill's expression, laws of nature were mere observational regularities "reduced to their simplest expression." His ideas were consequently taken up by conventionalists such as Pierre Duhem and – through Mach – by the logical positivists of the Vienna Circle, whose views differed from those of the older empiricists only with respect to the nature of logic and mathematics, which they regarded as *analytic* (in contradistinction to the synthetic *a posteriori* knowledge of empirical sciences). Thus in their pamphlet *The Scientific Conception of the World: The Vienna Circle* (written by Otto Neurath in 1929), one reads that vitalistic concepts in biology such as Reinke's dominants or Driesch's entelechies ought to be rejected as "metaphysical" since those concepts "do not satisfy the requirement of *reducibility to the given*".[31] Carnap's first attempt to formulate a programme for the unification of science, which was to use an *experiental* "reduction basis" (phenomenalism), was conceived in the same spirit. The subsequent reductionist programmes of the logical positivists which focused on inter-theory reduction (see the next section) allowed for theoretical terms to occur that were not directly reducible to "the given", but remained tributary to classical empiricism in that they required that reducing theories explain "at least the same domain" of phenomena (cf. also Popper with respect to successional theory replacement) as the reduced theories they had to replace, and preferably more.

(v) Next we can distinguish a meaning of reduction(ism) that differs completely from the preceding ones and is related to the philosophical problems surrounding the *foundations of mathematics*. Thus the logistic programme of unification of mathematics due to Russell and Whitehead is sometimes described in terms of the "reduction" of mathematics to logic.[32] I will not (and am unable to) discuss the question of the success of this programme here, which many consider as defunct by now. I only want to stress that "reduction" in this sense has nothing to do with reductionism as an explanatory stratagem in empirical science, since to say that one branch of mathematics could be "explained" by means of another branch seems entirely devoid of meaning, *even* if one acknowledges – as adherents of marxist epistemology do – that the analytic/synthetic distinction is not as clear-cut (cf. Quine) as logical positivists and certain other formalists would have it. (The analyses of the Geneva school in genetic epistemology are particularly instructive here).

I also want to remark that from the perspective of the *critique of ideology in contemporary science*, this subject is not necessarily as unworldly as it may seem at first glance. In fact, it has sometimes been argued that pre-Gödelian philosophy of mathematics and in particular Hilbert's programme should be considered as a grand "master plan", paving the way for a "supreme science of laws" (cf. Husserl's "nomologie" and the Cartesian "mathesis universalis"). The search for such an absolute and unshakable foundation of knowledge is then linked to attempts at self-legitimation of the powers that be by resorting to "mathematical discourse". Although the existing "analyses" of the role of (unified?) mathematical knowledge in ideological discourse – such as the one proposed by the "new philosopher" André Glucksmann[33] – cannot be taken at face value and are very superficial indeed, the problem itself seems important enough to be investigated in depth.

(vi) Finally, I think it is necessary to mark off one more type of reduction that differs considerably from the intertheoretical reductions highlighted in the analysis of the logical positivists, a type that was briefly mentioned by Vittorio Somenzi: *methodological reductionism*. Although more than one form of methodological reductionism can be distinguished, I think it is fair to say that what is essentially involved here is the phenomenon of "imperialist" intrusion of a paradigm (in a sense to be specified) that has actually (or presumably) been successful in one area of science into another area, the subject matter of which is far removed from the first. As an example, one could cite the introduction into economics of mechanistic models (and the mathematics appropriate to them) by Walras; or the systematic conquest of larger and larger parts of what had traditionally been considered as the subject matter of other social or humanistic sciences by the economic paradigm ("sociological economics", "economic analysis of rights", etc.) Roger Cavallo[25] has written some beautiful pages on this. Many forms of methodological reductionism went hand in hand with the promulgation of the *analytic* method which, together with the mechanistic explanations associated with it, has come to be seen by many through the centuries as "the *only* way to true knowledge". Moreover, Cavallo insists that the formalist spirit which accompanied the rise of modern algebra in the nineteenth century may well have contributed to the "extreme development of mathematics as a study independent of sense observations or any 'other reality than symbol manipulation' [an expression of De Morgan – W.C.] itself." He conjectures that a less analytic and more synthetic view of mathematics might have contributed to mathematical developments "more integrated with, attuned to, and suited for observations whose description was desired, including those from the social sciences" than those that have actually taken place. This in turn could have prevented the situation from occurring (which still exists today) in which empirically oriented studies are left to a large degree "to make do with what they have", or to consult the mathematician disenchanted with the purity of mathematics (Cavallo calls this attitude the "applied mathematics methodology") – a situation that makes in particular the social sciences vulnerable to "methodological imperialism".

With a view to recent developments in the philosophy of science, it can be argued that *methods are but one of the "exportable" elements* – apart from theories – *of a paradigm,* or a "disciplinary matrix" (Kuhn), or a "research programme" (Lakatos), *or whatever broad category one prefers to designate the unit of analysis which is at stake in scientific development.* Thus Darden and Maull, whose heterarchical concept of "interfield theories" is briefly discussed hereafter as an alternative to hierarchical reduction, have adopted the concept of a field – inspired by Toulmin and Dudley Shapere's work – to designate the unit involved in (intertheory) reductions. A *field* consists of many types of elements, such as a central problem, a domain (a set of "facts"), general explanatory factors and goals providing expectations as to how the problem is to be solved, techniques and methods, and (sometimes) concepts, laws and theories.[34] One could envisage all kinds of "migrations" of these elements, isolated or combined, from one field to another, which could be regarded as so many forms of reduction in analogy to the narrower form of methodological reduction discussed above. And one could study the impact of such migrations (positive or negative, etc.) on the field into which they are imported, the changes which they undergo themselves or which they induce in other elements of this field, and so on.

The logical-positivist programme of reduction as a case of an instructive failure

The reductionist programme, the historical development and decline of which are investigated schematically in this section, concerns *the unification and simplification of science as a "body of ordered knowledge"* (Carnap) *by means of the reductive elimination of higher-level theories, branches of science, or disciplines,* to the exclusion of other, non-explanatory forms of reduction. No attention will be paid to the pioneering attempts of certain logical positivists (Carnap and Neurath) to construct a unitary language of empirical science by defining the entire scientific vocabulary in terms of "sensationalistic" (experiental) predicates (phenomenalism) or in terms of the observable properties of physical "things" (physicalism). The unfeasibility of these endeavours was readily recognized, and there would be no point in repeating the arguments against them here. Instead, I will immediately turn to the proposals of J.H. Woodger[35] and Ernest Nagel,[36] which are usually characterized as attempts at *direct reduction* of one theory to another. Note from the outset that like other logical-positivist reconstructions, their analyses have been given a "linguistic turn" (i.e. they employ a "formal mode" instead of the usual "material mode of speech"). Thus ontological issues are systematically avoided as linguistic considerations become predominant.

(i) *Direct reduction.* Both Woodger's and Nagel's concern was primarily with "metascientific" or "foundational" problems in biology and neurology – in line with the Vienna Circle's contention that (vitalistic) organicism in biology as well as "psycho-vitalism" were the principal refuge of "metaphysical" thinking, rather than the social sciences, the concepts of which were allegedly "closer to direct perception" and therefore more reliable. None the less, the one example of a "successful" reduction that Nagel has elaborated at length, the reduction of thermodynamics to statistical mechanics, was taken from physics, like most of the logical positivists' case studies that were to follow. There are considerable differences between physical theories on the one hand and biological and psychological theories on the other, having to do not only with the differing degree to which these theories have been axiomatized (or are axiomatizable) but also, and even primarily, with the *distinct kinds of complexities involved*. It is not an exaggeration to say that a real "quantum jump" occurs with regard to complexity, if one moves away from physics to other domains. The favouring of examples taken from physics must thus, in retrospect, be seen as a fundamental limitation. In their defence, logical positivists have insisted (and continue to do so) that their strategy – starting from the more simple and eventually proceeding to the more complex – was the most natural one and, in fact, the only fruitful one to follow. In a restricted epistemological sense, this is probably true. Yet their unwarranted optimism as to the possibility of unqualified and unlimited generalization of the mechanistic paradigm makes their own work smack of a "metaphysical" tendency as well (the "negative", to put it loosely, of the vitalism they wanted to combat), as is evidenced by their flirtation with behaviourism lasting over the years.

Essentially, Woodger's view of reduction amounts to the following requirements:[37]

(A) The theoretical vocabulary of the reducing theory, T_1 (which is supposed to be an *adequate* theory) must be a proper subset of the theoretical vocabulary of T_2, i.e. the theory to be reduced. ("Gene" would be an example of a theoretical term from biological theory not in the lower theory, say organic chemistry).

(B) For every term P in the theoretical vocabulary of T_2 but not in that of T_1, there must be a "biconditional" (or *bridge-law*), i.e. a statement of law according to which every individual x that has the property P also has the property M, and inversely; such that

(1) M is in terms of the theoretical vocabulary of T_1, and

(2) the biconditional is "well established", i.e. a *tested statement*. (This distinguishes it from earlier, syntactically defined biconditionals leading to allegedly purely "epistemological reductions".)

(C) The translation of T_2 by means of the biconditionals *follows* from T_1. In Nagel's account, theories of *a branch at a given time* are reduced, but the formal mechanism remains the same. However condition (A) is supplemented with the stipulation "(vocabulary of) a branch B at time t", and similarly, in conditions (B2) and (C), T's are replaced by B's.

bı

One of the big problems with these definitions is the interpretation of conditions (B) and (C) and of their interrelationship. Nagel thought of (B), which he labelled the condition of *definability*, as a necessary but insufficient condition of (C), the condition of *derivability*; while others have subsequently argued that derivability was automatically assured. The details of these arguments need not bother us here.[38] I only note that as a result, it was recognized that, in Kemeny and Oppenheim's terms, "the essence of reduction cannot be understood by comparing only the two theories" for this could lead to odd reductions, resulting only in a more complicated picture that did not "explain" things any better; they concluded therefore that "we must bring in the observations".[39] *All hopes of constructing a paradigm of reduction which was "inter-theoretical" in the strictest sense of "theory", i.e. confined to the "organizational" aspect of science (as distinct from the "data" aspect), had melted.*

(ii) *Indirect reduction.* The amended reductionist programme that was proposed next was to become by far the most influential of all known proposals. (Therefore, alternative accounts such as those of Quine or Suppes will not be considered here.) The idea of an indirect reduction came down to three essential requirements:

(A) The vocabulary of the theory to be reduced, T_2, must contain terms not in the vocabulary of T_1, the reducing theory. (This is a weaker requirement than (A) in the preceding case, since the theoretical vocabularies of the respective levels are no longer considered as a set of "Chinese boxes"; e.g. not all physical terms are biological.)

(B) Any observational data explainable by T_2 must be explainable by T_1.

(C) T_1 must be at least as well systematized as T_2.

As in the case of direct reduction, this definition can trivially be specified to apply to branches of theories, the former being defined as the mere conjunction of the theories at a given moment in time.

The "vocabulary" involved here was taken to consist of *logical* (i.e. analytical) terms, *observational* terms and *theoretical* terms. The "theoretical vocabulary" was defined as that part of the extra-logical vocabulary which is allegedly *not* used in recording observations.[40]

I will not repeat here the arguments of critics à la Kuhn and Feyerabend who have singled out the weaknesses of the theoretical/observational distinction of logical positivism. The idea behind the concept of "systematization of a theory" was that to be non-trivial and to make sense with a view to the goals that are pursued in reduction — unification and simplification of the *existing* body of knowledge[41] — an acceptable *"ratio" of simplicity to explanatory power* must be guaranteed. As Oppenheim and Putnam had recognized, "T_1 is normally more complicated than T_2". But this they took to be "allowable, because the reducing theory normally explains more than the reduced theory". Plausible as this may seem, one cannot say that the new proposal was very useful. Indeed, the concepts of simplicity and explanatory power are both highly problematical. I have already stipulated this with respect to simplicity. As for explanatory power, a generally applicable account of it is not available either, which even vigorous defenders of orthodoxy now admit.[42]

(iii) *Micro-reduction*. The "masterpiece" of the logical-positivist construction of reductionism was obtained by adding an elementary *mereology* (i.e. a calculus of the part-whole relation) to the concept of indirect reduction. With each branch of science was associated a specific *universe of discourse*, and the objects in the universe of discourse of, say, B_2 were taken as wholes which possess a decomposition into proper parts, all of which belong to the universe of discourse of B_1. Consequently, the relation "micro-reduces" was considered as *transitive*. Therefore, micro-reductions could be said to have a *cumulative* character. In this way, *"progress" in science could be ascribed exhaustively to either an increase in "factual knowledge" or an improve-*

ment in the "body of theories". Explanation and prediction were regarded as semantically identical and only pragmatically different (according to the so-called "symmetry thesis"). *Micro-reductionist replacement of a B_2 by a B_1 was seen as a privileged case of theoretical progress, yielding not only a more unified and systematized body of theories, but guaranteeing at the same time a considerable "postulational" economy* (by eliminating higher-level-specific terms) *as well as the explanation of the reduced theories, and through them, a "deeper" understanding of the phenomena accounted for in the reduced theories*. (It was also thought that reduction could *correct* the now reduced theories.)

(iv) *Assessment*. What was wrong with this picture of reduction? Limitations of space prevent me from discussing in any detail the weaknesses to which I have already pointed. I think it is fair to say that *none* of the assumptions that had been made has remained unchallenged. Historical and empirical scrutiny have revealed an infinitely more complex picture of scientific practice that just won't fit this scheme. One cannot do justice to the present "cross-current" in a few sentences. Moreover, it is certainly premature to try to establish a constructive alternative on the basis of the new insights, one that would be general enough to function as an ideologically acceptable and workable substitute. Thus, reduction and reductionism may genuinely be said to be "in a period of interregnum" (Wimsatt). Yet at the risk of overly simplifying, I think we ought to take a stance on a number of crucial issues. To conclude this section, I propose to focus attention on the following points.

(i) *Negative implications*. My first point is rather straightforward. One shouldn't lose his/her time in trying to tackle one or more of the intricate problems previously associated with the label "reduction" by making the same failures all over again. Yet such cases of "relapse" are often encountered. To a large extent, this obviously has to do with the communication gap that exists between scientific practice and the meta-scientific (philosophical) reflection of it.

(ii) *Ontological commitment*. At first glance, there is something paradoxical in the fact that while marxists – aware of the dangers of reification – usually avoid all reference to ontology, the logical-positivist doctrine of reduction has pretended to discard or even to transcend all ontological questions, while *at the same time* providing a screen of legitimation for actual or potential ideological abusers. Of course it could be argued that the problem lies entirely within logical positivism, namely, in its fundamental ambiguity or inconsequentiality, its continuous moving back and forth between unrealistically high methodological and epistemological standards and opportunistic alliances. (As for instance with behaviourism, which, according to its own account, ought to be considered as reduction in the "wrong direction"). But even when this is true, the problem remains *whether or not one should appeal to the ontological ramifications of the materialist world outlook to impose a check on speculative constructions*, particularly with regard to the "natural hierarchy" of systems. Here, as in the case of epistemology, the principal danger seems to lie in ascribing an "absolute" character (now of "reality" instead of significance) to some privileged level, say, elementary particles. In this respect, the "end of *atomism*" expounded so brilliantly in the work of Ilya Prigogine[43] is to be seen as one of the biggest blows at the spirit of mereological reductionism. What are the precise implications here of our view as to the "material unity" of the world?

(iii) *Statistical explanation*. To the extent that statistical explanations become more and more preponderant in fields corresponding to "higher-level" domains of phenomena ("universes of discourse") the alleged explanatory strength of micro-reductionism becomes even more problematic than it always has been. The deductive-nomological paradigm of explanation on which it was based by definition applies only to deterministic laws. And hitherto, no acceptable (inductive) counterpart to the D-N scheme has been elaborated that can be said to apply accurately to statistical inference. (Note that the "micro-laws" so fundamental in the reductionist endeavour are often of a statistical nature too!)

In his *Theory and Evidence*, Clark Glymour[11] has made the claim that with respect to the interpretation of what constitutes an adequate explanation for a phenomenon, logical positivists, Bayesian probabilists *and* their historicist and relativist opponents have succumbed equally to a "radical epistemological holism" he takes to be unwarranted by practice. All three groups are convinced, he argues, that "whatever the illusions of practice may be, evidence can only bear on the entire body of our beliefs, and cannot be parcelled out here and there." To this, he opposes a more modest picture of scientific practice according to which scientists labour to fit selected pieces of evidence to selected pieces of theory, and vice versa.

If this account of the relation of theory and observational data proves right, the consequences for attempts at encompassing explanations by means of micro-reduction or other global mechanisms are formidable. Therefore, I think we will have to investigate carefully what a marxist perspective on matters of selective and statistical explanation would look like; e.g. how can objective and subjective probabilities be interrelated dialectically?

(iv) *"Fundamentality" revisited.* One aspect of the problem of considering reducing theories to be more "fundamental" or "basic" than reduced theories (in the epistemological sense) is related to the *asymmetrical* interconnection of the levels involved, which is realized by means of the bridge laws. Concepts of the reduced theory are imported into the reducing theory, but the reverse does not hold – so the classical account tells us. Case studies have revealed that there are actual instances of concepts originally belonging exclusively to the reducing theory that have to be incorporated in the theory one wants to reduce before reduction can possibly be envisaged. For example Steven Kimbrough[44] has recently argued that the concept of *chromosome* must be incorporated in (Mendelian) genetics if one is to be able to make Mendel's second law "law-like" in the sense required by classical reductionism. In the light of such evidence, the idea of an asymmetrical relationship between the levels involved may be questioned again.

(v) *Supervenience.* One of the most ingenious attempts to secure a number of essential features of reductionism hitherto is due to Jaegwon Kim, whose concept and theory of "supervenience" have been used subsequently by Terence Horgan to redefine the status of bridge laws and by Alexander Rosenberg to provide for the commensurability and "potential reducibility" of Mendelian to molecular genetics.[45] Essentially, supervenience amounts to a relationship of determination (dependence) between the families of properties, such that two phenomena alike with respect to the second group of properties are necessarily alike with respect to the first group *without* there being a relationship of definability or entailment between the two families. The category of supervenience was devised originally in axiology, the theory of moral and aesthetic valuation, to express the "consequentiality" of valuational properties upon naturalistic (descriptive) properties. Suppose we say that "St Francis was a good man" (the example is taken from the work of R.M. Hare). It seems logically impossible to state this and to maintain at the same time that another man could be conceived of who, placed in exactly the same circumstances and behaving in exactly the same way, differed from him in this respect only, that he was not a *good* man.

The concept of supervenience has been given precise meaning by means of "possible-world" semantics and applied to reduction problems. Thus bridge laws have been reinterpreted as being true in all possible "worlds" in which the reducing theory is true, *without necessarily reflecting attribute-identities*. Or it has been argued that, although the strict requirement of connectability between theories such as population genetics and molecular genetics may not be warranted in accordance with the traditional view, the theories are nevertheless *commensurable*, that their succession represents genuine progress and not only scientific "change", and that the relation between theories guarantees unification. More generally, supervenience is viewed as a regulative methodological principle corresponding to *mereological determinism* interpreted as a principle of *synchronic* determination (and complementary to the *diachronic* determination of *causal determinism*).

As far as I can see, the ideological and more general philosophical implications of this new, more sophisticated analysis of reduction are until now entirely left unexplored.

(iv) *Research strategy*. My last and most speculative point concerns our strategy for tackling mind-body problems. As the limitations of the reductionist programme are little by little overcome, it will become easier – so one might guess – to sort out those aspects of the mind-body cluster which are related to "message-in-circuit" characteristics.[46] This will probably imply that many of the disputes which are currently centred on issues such as identity or correspondence will have to be redefined in terms far removed from mereology, and in certain cases will be seen to have been based simply on the wrong kind of question. This will have important consequences for *localization fallacies* etc. On the other hand, many of the problems of biology (except of course for the all-pervading informational and "cognitive" aspects) will probably continue to be related to part-whole issues, albeit holistically reinterpreted.

Towards an alternative to reduction

What alternatives, then, could be proposed to reduction and reductionism? It should be noticed that according to a (growing) number of people, no surrogate is really needed. For not only has the old reductionist programme proved unable to deliver the goods; the very *rationale* for pursuing it has been dismissed as scientifically inadequate as well as ideologically suspect. The second point needs no elaboration here. As to the first, which has to do with the *functions* originally associated with reductionism, I think a balanced judgement is in order if we are not to throw out the baby with the bath-water. When taken very broadly, the idea that *some kind of unification and systematization is needed to compensate the alienating consequences of overspecialization* (which, as Cavallo insists, takes on more and more pathological forms) will have to be upheld. It may be conjectured that many of the dangers of current reductionism could be avoided in a radically *anti-hierarchical* mode of organization of science, defined not solely as a "body of ordered knowledge" but primarily as a social institution of which knowledge, its product, is but one– societally relevant– aspect.

Proposals that go in this direction have their roots in a reaction against the very deficiencies of reductionism as currently revealed. In this they differ from most "historical" alternatives to reductionist thinking such as von Bertalanffy's systems theory – which were less "technically" oriented but rather inspired by a general world view. I think that marxists can benefit from studying such new developments provided they are taken for just what they are: *preparatory work* that ought not to prevent us from going further.

On the other hand, *it cannot be excluded a priori* that in the light of the general redefinition of the place and function of philosophy of science vis-à-vis science and society which is now under way, *reductionism as a research strategy will itself be fundamentally reinterpreted*. The logical positivists tended to think that "nothing is to be lost, all to be gained" if a reductionist strategy is adopted. Both facets of this credo have been shown to be false. The shift from justification context to context of discovery which we currently witness may lead to a redefinition of the proper place of reduction in science as one type of *heuristics* (fruitful for generating new hypotheses in some contexts, inadequate and inconvenient in others) among many others. From this perspective, it is useful to observe that pro and con arguments concerning reductionism are used in at least two cases– evolutionary biology and cognitive psychology – as instruments in the theoretical disputes *within the science itself*, and are no longer confined to debates between philosophers. Needless to say, this is not only a challenging situation but also one which is likely to give rise to new and unforeseen dangers.

To finish, I would like to point once more to Darden and Maull's analysis of interfield theories, which represents an interesting broadening of the unit of analysis with which philosophers of science have traditionally been concerned. Instead of focusing on formalized single theories and on a single relation between them – derivational reduction– Darden and Maull explore the "virgin land" of encompassing scientific fields (in the sense defined earlier) and the interesting things that happen when two such fields are confronted. It appears that fields are not simply connected to each other but that new "inter-fields" are likely to be generated, the introduction of which is necessary, among other things, to answer questions which have arisen in a field but cannot be answered by the concepts, methods (etc.) of the field alone. I think the *network view of science* expounded here could profitably be coupled with a more sociologically oriented view of scientific *heteronomies*. This leaves me with the final impression that there is some hope for a bright future, after all – also with respect to the non-hierarchical organization of process and product of science.[47]

Notes and references

1. This formulation conflates Allan Muir's characterization of reductionism and a point Martin Barker has made about the distinction between explanation and description in his discussion of Steven Rose's views. There may be minor disagreement among the participants here, but we all seem to agree that reductionism is "the dominant *explanatory* mode in contemporary capitalist science" (meeting report draft).

2. "Kämpf um Gesundheit, die ärtzlichen Utopien", Frankfurt 1974, vol. 2, pp. 543-4.

3. I agree with Martin Barker that investigating the ideological nature of reductionist thinking and testing its truth are enterprises that should not (and cannot in practice) be separated. In the philosophy of science, this artificial gap has been cemented in the distinction between an (irrational) "context of discovery" and a "context of justification" (open to philosophical investigation and "reconstruction"), so dear to the Popperians. Such a view is not only inapt to cope with real scientific practice, as is more and more recognized nowadays (see Herbert A. Simon, *Models of Discovery*, Dordrecht 1977, pp. 326 ff. and my discussion of Wimsatt); it is also inherently dangerous, as it diverts attention from judging the (normative) (ir)rationality of ideologies and other "valuational" aspects of human action. On the other hand, it seems convenient to me to continue to distinguish the truth (or falsehood) of a theory in the sense of correspondence between "intellect" and "fact" from the "self-realization" of a theory (which may be false in this first sense) by means of a mechanism of "self-fulfilling prophecy", which "makes" this theory true in a larger Hegelian sense by adapting reality to it. (This second and rather obscure concept of truth could be clarified by introducing the time dimension into the first concept.)

4. I see no compelling reason why the left should join with "anti-theoricist" philosophers and historiographers of science whose position is in the last analysis found to be based on an unwar-

ranted belief in the superiority of "self-regulative" mechanisms in science.

5. Cf. Martin Barker's definition of reductionism as "the epistemological proposal and practice of seeking explanatory accounts of phenomena in terms of some fundamental principles". His further qualification ("in which complex phenomena are understood in terms of basic parts out of which they are composed") is more restrictive; it is confined to those forms of reduction which are mereological (such as Oppenheim and Putnam's "micro-reduction") and which must be distinguished from other forms of reduction, e.g. causal reduction. Mereological reductionism involves a form of cross-level determination which is not necessarily of the causal type.

6. A. Wilden, *System and Structure*, London 1980, p. 93.

7. See W.C. Wimsatt, "Reduction and reductionism", in P.D. Asquith and H.E. Kyburg, Jr. (eds.), *Current Research in Philosophy of Science*, East Lansing 1979, pp. 355 ff. This is one of the most informative discussions of reductionism – written from a post-positivist perspective – I have come across, one we ought to analyse in detail for all its intriguing implications.

8. "The heritage from logical positivism: a reassessment", in F. Suppe and P. Asquith (eds.), *PSA 1976*, East Lansing 1977, vol. 2, p. 254.

9. See L. Apostel, *Logique et dialectique*, Ghent 1979. Dialectics could be clarified by means of dynamic logics describing the transition of one system of axioms to another (as studied by the Piaget group in Geneva), non-monotonic logic (as elaborated in artificial intelligence), and, of course, by a number of developments in the social sciences as well as in biology.

10. By "phenomena", I understand processes as well as things, properties and relations. Much of the neopositivists' work on reduction is hampered by a heavy "structuralist" bias. Though their concept of form of order (*Ordnungsform*) did not exclude a process-view of reality, they have in practice often only attended to (spatially extended) "things". Cf. also Chris Sinha's point about subject and predicate in *Towards a Liberatory Biology* (companion volume).

11. C. Glymour, *Theory and Evidence*, Princeton 1980, p. 43. An instructive discussion of some of the problems involved in attempts at a broader definition of (psycho- and socio-linguistic) complexity may be found in R. Bartsch, "Gibt es einen sinnvollen begriff von linguistischer komplexität?", in *Zeitschrift für germanistische linguistik* no. 1, 1973, pp. 6-31.

12. "Science and Simplicity", Voice of America Broadcast.

13. That Marx relied too heavily on a pragmatic notion of truth (see especially the *Theses on Feuerbach*) is one of the very few points on which I would agree with Popper. On the relationship between semantical truth (the correspondence principle of realism) and pragmatical truth, see L. Apostel, *Theory of Knowledge and Science Policy*, Ghent, September 1980, pp. 18 ff; and N. Rescher, *Methodological Pragmatism*, Oxford 1977. Rescher's "move" consists in a shift from what he labels "thesis pragmatism" ("a proposition is said to be true if its adoption is maximally success-promoting") to "methodological pragmatism" ("a proposition is said to be acceptable, i.e. to qualify as true, if it conforms to an epistemically warranted criterion or method, and a method is warranted if its adoption as a generic principle for propositional acceptance is maximally success-promoting"). The latter he takes to be superior in that it rules out "gratuitous success" (false beliefs can sometimes be practically or "survivalistically" efficacious, while true beliefs can be counterproductive in these ways). This shift is reminiscent of the transition from the elder act utilitarianism in moral philosophy to "rule utilitarianism", the view that the criterion of utility must be applied not to individual acts but to the general rules governing these acts – a transition that yields a less unworldly theory of morality. I believe a marxist epistemology could benefit from adopting Rescher's basic idea.

14. "How complex are complex systems?", in Suppe and Asquith, op. cit. p. 508.

15. See especially his "The architecture of complexity", in *The Sciences of the Artificial*, Cambridge, Mass. 1969.

16. A chapter of my Ph.D. thesis will be devoted to these topics.

17. "How complex . . .", p. 510.

18. Marxists are often unwilling to grant that a well elaborated evolutionary epistemology could be compatible with their endeavour and have rightly pointed to the ideological dangers inherent in existing theories (e.g. those of Popper, Toulmin or Donald Campbell). Conversely, such a position is subject to a kind of Cartesian dualism entirely absent in Marx, to name but him. Many of the deficiencies of existing evolutionary epistemologies seem to have to do with precisely the defaults of neo-Darwinist evolutionary biology as described by Levins-Lewontin, Ho-Saunders and others. But to defend this thesis I would need the space of another paper.

19. Cf. Wilden, op. cit. p. 349 (on long-range survival value and on the "fear of survival", including the fear of radical socio-economic reorganization, which in Western society seems to have become greater "than the fear of death itself").

20. See H. R. Maturana and F. J. Varela, *Autopoiesis and Cognition*, Dordrecht 1980.

21. op. cit. p. 93.

22. The traditional view had no place for such pragmatic considerations; thus, it failed to capture an essential characteristic of the dynamics of science. See in particular C.G. Hempel, *Aspects of Scientific Explanation and Other Essays in the Philosophy of Science*, New York 1965.

23. *Foresight and Understanding*, New York-Evanston 1963, p. 63.

24. *The Nature of Explanation*, Cambridge 1967.

25. E.g. R. Cavallo, *The Role of Systems Methodology in Social Science Research*, Boston 1979, p. 14.

26. *Method, Model and Matter*, Dordrecht 1973. Cf. Cavallo, ibid.

27. Thus, the syntactic meta-language Hilbert introduces in his *Foundations of Mathematics* is "completely empirical" in that it contains only signs. (I owe this point to Leo Apostel.)

28. I agree with Steven Rose that one should be careful with metaphors deriving from cybernetics, information theory and artificial intelligence. Nevertheless I think that with a view of the general state of our knowledge of the human mind, we would be wrong to try to do without them.

29. F. Geyer's recent study of alienation (*Alienation Theories, A General Systems Approach*, Oxford 1980) is in part tributary to Luhmann's analysis of complexity reduction, which to date remains largely unknown in the Anglo-Saxon world. An essential difference between Luhmann's view on complexity reduction and Simon's theory (or, more generally, reduction in the first sense) is that the former's concept of environment (cf. "Welt", as related to Husserl's "Horizont") is to be seen as itself a social construction. Thus complexity never refers to a "state of being" (*Seinszustand*) but always to a relationship between system and environment.

30. See R. Harré, "History of Philosophy of Science", in the *Encyclopedia of Philosophy*.

31. *Wissenschaftliche Weltauffassung: Der Wiener Kreis*, Dordrecht 1973, p. 16.

32. H. Putnam, "Philosophy of Mathematics", in Asquith and Kyburg, *op cit.*, pp. 386-98.

33. *Les Maîtres Penseurs*, Paris 1977, pp. 275-7.

34. "Interfield theories", in *Philosophy of Science*, no. 44, March 1977, pp. 43-64.

35. *Biology and Language*, Cambridge, Mass. 1952, pp. 271-2 and 336-8.

36. "Mechanistic explanation and organismic biology", in *Philosophy and Phenomenological Research*, no. 11, March 1951. In Chaps. 11 and 12 of his *The Structure of Science* (London 1961), Nagel has updated his views in the light of later developments.

37. This interpretation is based on the accounts of K. Schaffner ("Approaches to reduction", in *Philosophy of Science*, no. 34 1967, pp. 137-47) and J. G. Kemeny and P. Oppenheim, "On Reduction", in *Philosophical Studies*, no. 7, 1956, pp. 9-19.

38. Nagel's reason for regarding (C) as a condition independent of (B) seems to have been the consideration that T *must have been actually written down* to be recognized as a theory of a branch at time t. His critics felt that such "pragmatic" considerations should not be allowed to creep into logical analyses!

39. op. cit. p. 13.

40. op. cit. and P. Oppenheim and H. Putnam, "Unity of science as a working hypothesis", in H. Feigl *et al.*, *Minnesota Studies in the Philosophy of Science*, vol. 2, Minneapolis 1958, pp. 3-37.

41. This follows immediately from the nature of logical *re*constructions. Talking of "potential reduction" or "reducibility in principle" is rather futile, as Woodger has already insisted. Wimsatt (op. cit., pp. 357 ff.) shows convincingly that "in principle" accounts are usually extremely difficult to falsify. Moreover, "in principle" feasibility is not a "trivial", "empirical question" (in this case, related to the state of development of the sciences), since it may involve *in*feasibility to us as human calculators (ibid.).

42. Glymour, op. cit. pp. 42-3.

43. See in particular "La fin de l'atomisme", *Acad. Roy. Belg. Bull. Acad. Cl. Sci.*, 5th series, vol. 55, no. 12, 1969, pp. 1110-17.

44. "On the reduction of genetics to molecular biology", in *Philosophy of Science*, no. 46, 1979, pp. 389-406.

45. "Supervenience and nomological incommensurables", in *American Philosophical Quarterly*, no. 15, 1978, pp. 149-56; "Supervenient bridge laws", in *Philosophy of Science*, no. 45, 1978, pp. 227-49; "The supervenience of biological concepts", ibid. pp. 368-86.

46. See A. Kuhn, "Dualism reconstructed", in *General Systems*, no. 21, 1977, pp. 91-7.

47. See H. Nowotny, "Heterarchies, hierarchies and the study of scientific knowledge", in W. Callebaut *et al.* (eds.), *Theory of Knowledge and Science Policy*, Ghent 1979, vol. 1, pp. 536-50.

Notes on contributors

to Against Biological Determinism *and* Towards a Liberatory Biology

Enrico Alleva is Research Associate at the Department of Pharmacology (Psychopharmacology Section), Istituto Superiore di Sanità, Rome.

Martin Barker is Lecturer in Philosophy in the Department of Humanities at Bristol Polytechnic and is a member of the editorial board of *Radical Philosophy*.

Patrick Bateson is Reader in Animal Behaviour at the Sub-Department of Animal Behaviour, Madingley, Cambridge (University of Cambridge Department of Zoology).

Giorgio Bignami is Senior Research Scientist in the Department of Pharmacology (Psychopharmacology Section), Istituto Superiore di Sanità, Rome, working on the mechanisms underlying behavioural changes induced by drugs.

Lynda Birke researches on hormones and behaviour in the Biology Department at the Open University, Milton Keynes.

Werner Callebaut is researching on hierarchical organization at the Faculteit van de Letteren en Wijsbegeerte, Rijksuniversiteit, Ghent, Belgium.

Bea Chorover is a Family Therapist at the Brookline Mental Health Center, Brookline, Massachusetts.

Stephan Chorover is Professor of Psychology at Massachusetts Institute of Technology, Cambridge, Massachusetts.

Jonathan Cooke researches on early embryonic development in the Division of Developmental Biology at the National Institute for Medical Research, London.

Bruno D'Udine is researcher in genetics and ontogeny of behaviour at the Istituto di Psicologia e Psicofarmacologia, National Research Council of Italy, Rome.

Lauro Galzigna is Professor of Biochemistry at the University of Padova, Italy, and has formerly worked in the USA, France, Germany and Kenya.

Giacomo Gava teaches Philosophy at the Istituto di Storia della Filosofia at the University of Padova, and has worked mainly on the philosophy of science, psychology and neurophysiology.

Brian Goodwin researches on developmental biology and teaches in the Department of Biology at the University of Sussex, Brighton.

Mae-Wan Ho is Lecturer in Biology at the Open University, Milton Keynes, and researches on genetics and evolution.

Ruth Hubbard is Professor of Biology at Harvard University, Cambridge, Massachusetts, and has been active in the women's movement, examining how male domination of the scientific professions has affected their theoretical content and social practices.

Allan Muir is Lecturer in Mathematics at the City University, London, and researches on cybernetics and automata.

Lesley Rogers is Lecturer in Pharmacology at Monash University, Australia, researching in the field of brain and behaviour.

Hilary Rose is Professor of Applied Social Studies at the University of Bradford. She has been an activist in the radical science movement and is presently engaged on a feminist critique of the masculine character of scientific knowledge.

Steven Rose is Professor of Biology at the Open University, Milton Keynes, and researches and writes on brain mechanisms and learning, and science and its social relations.

Peter Saunders is Lecturer in Mathematics at Queen Elizabeth College, London University.

Claudio Scazzocchio teaches in the Department of Biology, University of Essex, Colchester.

Chris Sinha researches on the problems of language, psychology and education and works at the Hester Adrian Research Centre for the study of learning processes in the mentally handicapped, at Manchester University.

Janna Thompson works in the Department of Philosophy, La Trobe University, Australia.

Gerry Webster has researched on the problem of biological form and the history and philosophy of science, and teaches in the Department of Biology at the University of Sussex, Brighton.

Index